THE NEW BIRDER'S GUIDE TO

BIRDS OF NORTH AMERICA

THE NEW BIRDER'S GUIDE TO

BIRDS OF NORTH AMERICA

BILL THOMPSON III

ILLUSTRATIONS BY
JULIE ZICKEFOOSE
AND MICHAEL DIGIORGIO

HOUGHTON MIFFLIN HARCOURT
2014

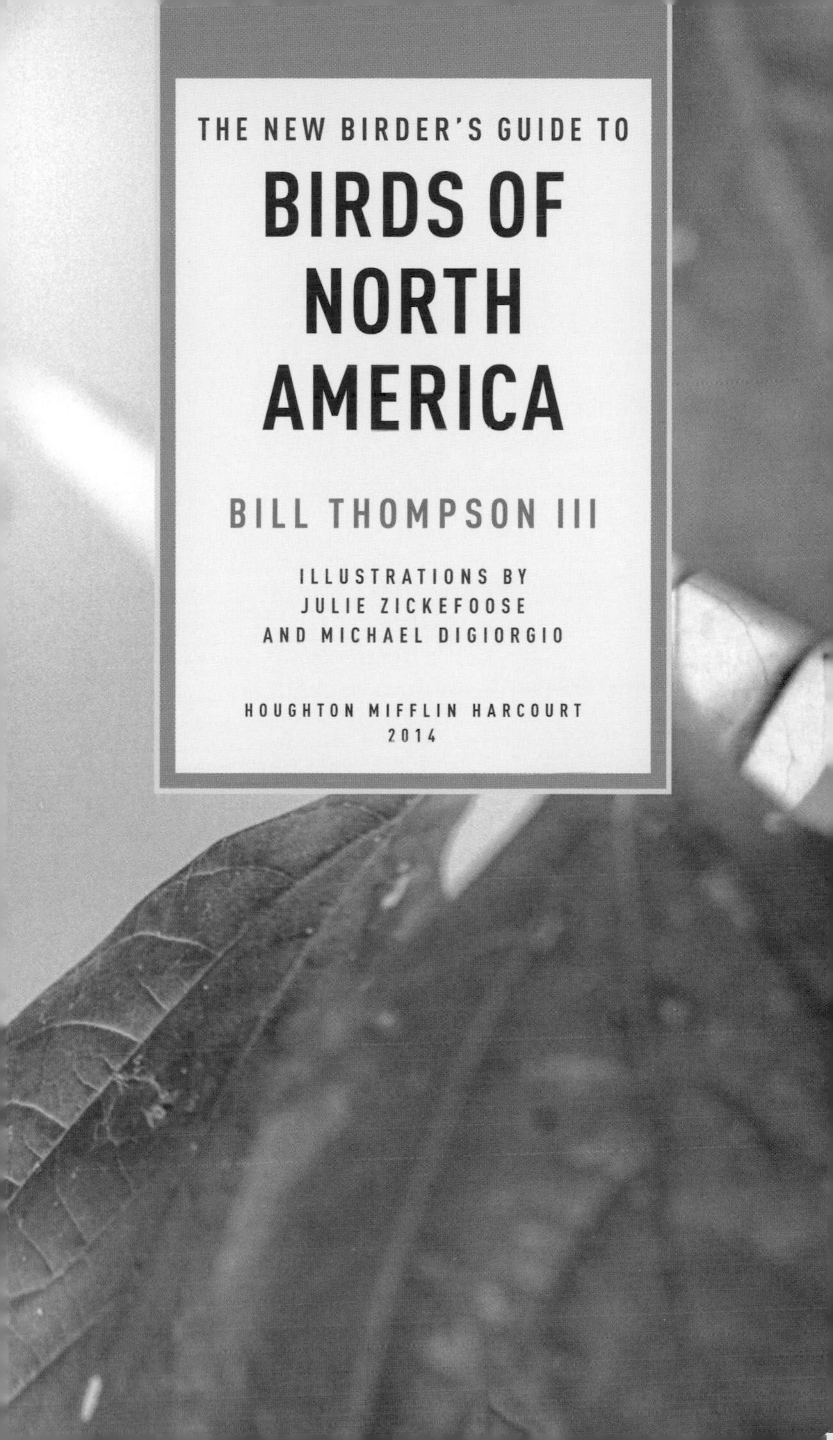

For information about permission to reproduce selections
from this book, write to Permissions, Houghton Mifflin Harcourt,
215 Park Avenue South, New York, New York 10003.

www.hmhbooks.com

Library of Congress Cataloging-in-Publication Data is available.
ISBN 978-0-544-07047-9

Printed in China

SCP 10 9 8 7 6 5 4 3 2 1

To all the birding mentors out there who are "paying it forward" by taking the time to help new birders enjoy this fabulous hobby of bird watching.

PHOTO CREDITS

Glenn Bartley: 143, 177, 259 right, 264

Cliff Beittel: 52, 66 left, 94 left, 96, 97, 100, 102 left, 106, 117, 119 left, 127, 128 left, 129 left, 134 left, 137 left, 138, 155, 156, 157, 162, 170, 172, 193 left, 200 right, 202 left, 209, 215, 232 left, 233, 244, 246, 248, 265, 269, 272 right, 310 right, 319 left, 327 right, 334 left

Wendy Clark: 15

David Dvorak Jr.: 115, 141 right, 227 left, 257 left

Brian Henry: 91, 165, 175 right, 245, 258 left, 258 right, 272 left, 277 right, 282 right, 325 bottom

Greg Lasley: 161, 243 left, 247

Jerry Liguori: 105 left, 116 top, 118 right

Maslowski Productions: 58, 72 right, 80 left, 104 left, 135 right, 144 left, 167, 168, 180, 181 left, **181** right, 188 right, 203, 218, 222, 227 right, 230, 238, 255, 256 right, 281 right, 284 left, 293 left, 306 right, 312 right, 316 left, 317 right, 319 right, 324 right, 332 right, 342

Garth McElroy: 256 left, 281 left, 282 left, 283 left, 286, 287 left, 287 right

Charles Melton: 51 top, 65, 68 left, 68 right, 71 left, 71 right, 79 left, 83 right, 88, 92, 94 right, 95 bottom, 101, 103 right, 105 right, 107, 108 left, 108 right, 125, 128 right, 136 right, 140, 142 left, 144 right, 164, 171, 179, 182 left, 182 right, 183 left, 183 right, 184 right, 185 left, 185 right, 186 top right, 186 bottom right, 187 left, 187 right, 191, 195, 199, 210 left, 210 right, 214, 236, 237, 250, 260 right, 261, 266, 274, 280, 290, 292 right, 294 left, 294 right, 295, 297 left, 297 right, 298, 299 right, 309 right, 313 left, 313 right, 315 left, 315 right, 316 right, 317 left, 318 right, 331 right, 333 left, 333 right, 335, 339 left, 339 right, 344 left, 344 right, 345 left, 345 right

Gary Meszaros: 109 left, 109 right, 166 right, 178, 204, 257 right

Ellen Michaels: 193

Arthur Morris/Birds As Art: 20, 50, 51 bottom, 56, 62, 81 left, 82, 87, 90 right, 93, 95 top, 102 right, 103 left, 104 right, 110 right, 112, 114 right, 123, 126, 130 left, 131, 133, 134 right, 135 left, 136 left, 141 left, 147 right, 150, 152 left, 152 right, 153 right, 159, 166 left, 276, 278 left, 283 right, 307 left, 311 right, 314 left

Tom Murray: 80 right

Marie Read: 53, 55, 57, 60 right, 61, 63 left, 67 left, 69, 76 left, 76 right, 77 left, 77 right, 86 right, 89 right, 98 left, 98 right, 110 left, 111, 120, 124, 145, 146, 158, 160, 176, 190 left, 190 right, 192, 194 left, 194 right, 196 right, 198 right, 202 right, 207, 213, 216, 219, 221, 226, 231 left, 231 right, 234, 241, 242, 253, 263 left, 270, 279, 289 right, 293 right, 296 left, 302, 305 top, 306 left, 309 left, 310 left, 312 left, 314 right, 321 right, 325 top, 326, 330 right, 332 left, 338 left, 341 left, 341 right

© **robertmccaw.com:** 54, 59 left, 59 right, 60 left, 63 right, 64 left, 64 right, 66 right, 67 right, 70 left, 70 right, 72 left, 73, 74 left, 74 right, 75 left, 75 right, 78, 79 right, 81 right, 83 left, 84, 85 left, 85 right, 86 left, 89 left, 90 left, 99, 113, 114 left, 116 bottom, 118 left, 119 right, 121, 122, 129 right, 130 right, 132, 137 right, 139, 142 right, 147 left, 148 left, 148 right, 149, 151 left, 151 right, 153 left, 163, 169, 173, 174 left, 174 right, 184 left, 186 left, 196 right, 197, 198 left, 200 left, 201 left, 201 right, 205, 206, 208, 211, 212, 217, 220, 223, 224, 225, 228, 229, 232 right, 235, 239, 240, 249, 251, 252, 254, 259 left, 260 left, 262, 263 right, 267, 268, 271 left, 271 right, 273 right, 275 left, 275 right, 277 left, 278 right, 284 right, 285 right, 288, 289 left, 291, 292 left, 299 left, 300, 301, 303, 304, 305 bottom, 307 right, 308, 311 left, 318 left, 320 left, 322 left, 322 right, 323 left, 323 right, 324 left, 327 left, 328 left, 328 right, 329 left, 329 right, 330 left, 331 left, 334 right, 336 left, 336 right, 337 left, 337 right, 338 right, 340 left, 340 right, 343 left, 343 right, 346, 347 left, 347 right

Barth Schorre: 273 left, 285 left

Brian Small: 154, 189, 243 right, 321 right

Bill Thompson III: 10, 24, 27, 39, 42, 43, 49, 320 right

Julie Zickefoose: 32–33, 175 left, 188 left

CONTENTS

WHAT IS BIRDING?

Birding, or bird watching (the two terms mean essentially the same thing), is just what it sounds like. The activity of identifying a bird involves seeing a bird, looking at it closely through binoculars, and then trying to match what you see to a picture in a field guide (like the one you are currently reading).

Birds are divided into all sorts of categories as determined by bird scientists, or ornithologists. This organization of birds is called taxonomy, and it is based on the similarities (or relatedness) of birds. The two categories that are most important to the bird watcher are bird families and bird species. A bird species is defined as a distinct group of identical or very similar individuals that can successfully produce offspring. A bird family is a group of related bird species. The Downy Woodpecker is a bird species and is a member of the woodpecker family, which includes 23 different related species in North America.

Some birders enjoy correctly identifying the birds they see. "*That* bird is an Eastern Wood-Pewee! I know that bird as sure as I know my next-door neighbor!" Other birders just want to see the birds and enjoy them without worrying about putting a name on each bird.

Birds, because they have wings and tend to use them, can show up anywhere at any time. Every little scrap of habitat can contain birds. When you stop to gas up the car on a trip, there will be birds. Coming home from work, you'll hear birds in the trees and see them flying overhead. Birds are everywhere. Bird watchers know this and take advantage of the chances to see them and enjoy them.

◀ *It's always more fun (and more beneficial) to go birding with a small group.*

GETTING STARTED IN BIRD WATCHING

Many birders start out looking at birds in the backyard—maybe those species visiting the bird feeders or birdbath. After becoming familiar with most of the birds in the backyard, a bird watcher might venture out to look for birds in a local park or along a nearby stream or lake.

Once you get interested in birds, don't be surprised if you find that you'll want to see more and different birds. This usually means that you'll need to travel a bit. And there's a whole world of birding adventure out there just waiting for you. More than 800 bird species are regularly found in the United States and Canada. Worldwide, there are more than 10,000 bird species. So if you start birding today, it's going to be a long time before you run out of birds to see and enjoy.

Birding is a hobby that is inexpensive and easy to do. You can watch birds anywhere and at almost any time. For many of us who enjoy watching birds, we can trace our interest to a single encounter that sparked our imagination. This is our "spark" bird. My spark bird was a Snowy Owl that drifted into the giant oak tree in front of our house in Pella, Iowa, on a cold November day. I was six years old and was helping my parents rake the leaves off the front yard when a large white bird caught my eye. I was stunned at the bird's size, its clean white plumage, and its mysterious appearance. I knew it was an owl, but what kind of owl? And weren't owls creatures of the night? I ran inside the house for our very basic field guide. There it was: Snowy Owl! I devoured the short description: *A huge, mostly white bird of the Arctic tundra. Ventures southward in some years when food shortages and severe weather force it to move.* I spent the next few days carrying that guide around our large backyard, looking at birds and then trying to find them in the guide. For me this new activity was just like a nature scavenger hunt and I've been pursuing birds ever since.

Once you see your spark bird, or even if you have yet to see it, the best way to get good at birding is to watch birds whenever you can. Even better, join up with a friend or a local bird club. You will see more birds and learn more about them if you go birding with others. It's simple: more eyes, ears, and brains mean you see more birds and have more fun.

But Is Birding Cool?

Birding is one of North America's fastest-growing and most popular hobbies—there are as many as 44 million bird watchers in the United States. Back in 1978, when my family began publishing *Bird Watcher's Digest* from our living room, bird watching was still considered a little bit odd. In fact, if you're over the age of 40, chances are you'll remember the iconic image of a birder in those days: Miss Jane Hathaway from *The Beverly Hillbillies*. She was nervous, snooty, and nerdy: a lethal combination. And an image that was pretty unfair and inaccurate in its depiction of bird watchers. Today things are different. Birding is no longer considered a hobby for little old ladies and absent-minded professors in funny hats. There are former presidents, rock stars, movie stars, and millions of regular folks like you and me who enjoy watching birds. Everyone knows at least one bird watcher among his or her friends and family. Once your friends realize that you know something about birds, they'll ask you all kinds of questions and will share bird sightings with you. Don't let anyone else's perception of birding bother you. If you like to watch birds, then, as one famous company advises, *Just do it.*

WHY WATCH BIRDS?

There's no question that birds themselves are cool and amazing. Birds have inspired human beings for thousands of years. Why? There are several reasons. Birds can fly whenever they want, wherever they want to go. We humans figured out how to fly only about 100 years ago. Birds are amazing and beautiful creatures—they have brilliant plumage and may change their colors seasonally. Many birds are master musicians, singing beautiful, complex songs. They possess impressive physical abilities—hovering, flying at high speeds, and surviving extreme weather and the rigors of long migration flights. In short, we admire them because they inspire us. This makes us want to know them better and bring them closer to us. We accomplish this by attracting them to our backyards and gardens, by going to find birds where they hang out, and by using optics to see them more clearly.

Basic Gear

If you're just starting out as a birder, all you really need to enjoy watching birds are two basic tools—binoculars and a field guide. The binoculars (often referred to as "binocs" or "bins" by bird watchers) help you see the bird better. The field guide helps you identify what it is you're seeing. Of course there's lots of other gear and many gadgets that you may add to your arsenal, but it all starts with binoculars and a field guide.

BINOCULARS: Binoculars are like two miniature connected telescopes that enlarge distant birds so that we can see them well enough to identify them.

You may be able to borrow optics from a friend or family member, but if your interest in birding takes off, you'll certainly want to have your own binoculars to use anytime you wish. (Try dropping hints to your family and friends: "Gee, I sure hope the Birthday Elves will bring me some good binoculars for birding!") Fortunately, a decent pair of get-you-started binoculars costs less than $100. And some really nice binoculars (even used ones) cost just a bit more.

The magnification powers that are commonly used for bird watching are 7x, 8x, and 10x. Power is always the first number listed in a binoculars description, as in 8x42. The second number refers to the diameter of the objective lens (the big end) of the binoculars, measured in millimeters. The bigger the second number, the brighter the view presented to your eye. In general, for bird-watching binoculars the first number should be between 7 and 10, and the second number should be between 30 and 45. A good place to start is with an 8x42 binocular.

Try to find binoculars that are easy to use. Make sure they are comfortable to hold (not too large or heavy), fit your eye spacing, and focus easily, giving you a clear image. Every set of eyes is different, so don't settle for binoculars that don't feel right. Over time, you will become superskilled at using your binocs, and that's when they will feel like an extension of your own eyes. You won't even notice that you're using them.

FIELD GUIDE: When choosing a field guide, it's helpful to know what type of birding you'll be doing and where you plan to do it. If nearly all of your bird watching will be done at home, you might want to get a basic field guide to the backyard birds of your region, or at least a field guide that limits its scope to your half of the continent. Many field guides are offered in eastern (east of the Rocky Mountains) and western (west of the Great Plains) versions. These geographically limited formats include

PARTS OF A BIRD WATCHER

To get the most out of your bird-watching adventures, you'll want to have the right gear. In birding, like in other sports, the emphasis is on function more than fashion. Good binoculars and a field guide are essential. The rest of the elements shown on our birder-model (above) are for convenience and comfort.

only those birds that are commonly found in that part of the continent, rather than continent-wide guides, which include more than 800 North American bird species. Choose a field guide that is appropriate for you, and you'll save a lot of searching time. And that's time that can be better spent looking at birds! This Field Guide is intended to be a good starter book as you begin watching birds. It contains 300 of the most common and often encountered birds of North America. Other field guides can be found at bookstores, in nature centers, and on the Internet. See the Resources section at the end of this book for suggestions.

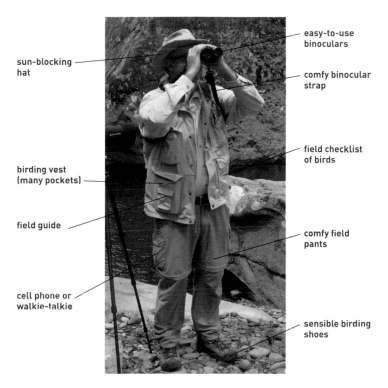

OPTICS: Having binoculars around your neck, ready for use at a moment's notice, is akin to having a superpower. Your binoculars magnify the birds you are trying to see, making them easier to identify. How amazing is it that we can magnify our vision by eight or ten times in an instant? Despite this, bird watchers of all levels of experience and expertise often neglect to carry their binoculars with them when they venture outside. I missed a great look at a passing flock of shorebirds—a rarity on the dry Ohio ridgetop farm where I live—because I left my binoculars on the table inside our front door. What were they? Some sort of sandpiper, but I'll never know for sure because they were gone before I could get my binoculars. There's nothing I like more than adding a new species to our yard list (which currently stands at 188 species). Now I strap on my binoculars when I'm taking the ten steps to toss out the kitchen compost. Even at night!

FIELD GUIDE POUCH: When I'm out guiding I like to carry my field guide in a pouch that belts around my waist. This keeps the guide safe from the elements and handy for quick reference. I recommend this.

CHECKLIST/NOTEPAD: Keeping a list of birds seen on an outing is a huge part of the fun for me. I always have a checklist along and ask someone (usually my daughter, Phoebe) to be the list keeper. These days, lots of birders keep records on their smartphones or digital tablets and share them via social media and citizen-science networks. I love that this is happening, but I can't bear to stare into my phone to update a digital checklist when I could be scanning for birds. By the way, if you lack an actual checklist, a pencil and the back of an old oil-change receipt works just as well.

PROPER FOOTWEAR: I ruined our honeymoon because I lacked the proper footwear. I was barefoot when I accidentally fell from a dock into the warm waters off the coast of Belize. Unfortunately, I caught my smallest left toe on the edge of the dock. This was 12 hours into our honeymoon (which was also a birding trip we were helping to guide—we were poor newlyweds). The rest of our honeymoon went like this: boat, doctor, tetanus shot, sedatives, strange hotel, wheelchair, airport, home to snowy Ohio. Surprisingly, Julie did not ask to annul our union on the spot. She never lets me forget it, however. So, hear me on this: match your footwear to the type of birding you'll be doing. I tend to go for toe and ankle support these days—no wonder. When I travel on extended overseas trips, I always take three different types of shoes: hiking boot or

FIELD GUIDE

My wife, Julie, and I were birding with our infant daughter Phoebe at a spot in southeastern Ohio known as The Wilds. We spied an unfamiliar female duck in among the hundreds of gadwall, mallards, ring-necked ducks, and lesser scaup. "Go get the field guide out of the car," Julie said. "Which bag did you put it in?" I replied. "I thought *you* packed the guide," she countered. We both looked at Phoebe. She was far too young to blame for this mistake. Julie made a rough sketch of the bird using a felt-tip marker and the back of an envelope. It was enough to help us later on, when we got home to our forgotten field guide, to determine that this was likely a female garganey. Our joy was somewhat muted, and then turned to disappointment, when a few days later the Ohio rare bird report was abuzz with sightings of a female garganey at a pond a half-mile from where we'd been. *Always take a field guide along.* (It doesn't have to be *this* field guide, though I'd be pleased if it were.)

waterproof wellies (depending on the climate and setting), tennis shoes that I won't mind ruining in the mud and leaving behind, and sandals or flip-flops to give my aching feet a rest from the first two.

HEADGEAR: If you're middle-aged like me, you'll recall the days when everyone talked about *suntan lotion* instead of *sunscreen.* Many of us suffered severe sunburns in our younger days, before we knew that the glowing yellow-orange orb in the daytime sky could fry us like bacon. I'm pretty sure I'm helping to put my dermatologist's kids through college with all my office visit fees. Needless to say, I take precautions to protect myself from further sun damage, starting with *always* wearing a hat when I am outside. Lately I've even begun wearing the wide-brimmed, white bush hats that so many nature enthusiasts wear. They are comfy, cool (temperature-wise) in a breeze, heat-reflective, and they keep the direct sun off my face, ears, and neck, even if they lessen your dignity somewhat. I don't really like how I look in these hats, but I wear them nonetheless. After all, I'd much rather sacrifice a bit of dignity than another hunk of skin.

SUNSCREEN AND LIP BALM: "If it ain't at least 30 SPF, you're just wasting your time," or so one older gent once told me as we were smearing white cream all over our faces and arms. I figured he'd lived that long because he was wise, so I've heeded his advice ever since. Despite all the creamy and clothing coverage, I still color up when I spend a sunny day outside birding. Notice I say "color up." That's because the only part of me that gets tan are my millions—perhaps billions—of freckles. They get dark. The non-freckled part of me just burns. And I'm not one of those conspiracy theorists who believes that moisturizing lip balm actually makes your lips more dry and chapped. I dislike chapped lips more than I dislike the Cincinnati Reds. Here's something I do believe in: smooth, unchapped lips, and an eventual world championship for my beloved Pittsburgh Pirates.

CELL PHONE, CAR CHARGER: In *Bird Watching for Dummies*, the first book I ever wrote, I recommended that bird watchers venturing afield should stash a quarter or two inside their field guide pouch in case they needed to make an emergency call from a pay phone. Right now you might be asking yourself one of the following questions:

What's a pay phone?

When did phone calls cost only a quarter?

I wonder what dinosaur species Bill saw while he was out birding back then.

These days, we all have a handy cell phone with which we can call for help, order pizza, or update our Facebook status about the birds we are seeing. But a cell phone lasts only as long as the life of its battery, and if you're a constant Tweeter, text messager, or Siri questioner, your phone will die before the day is over. For this reason I keep a charging cord and car charger in my birding van at all times. In fact I've labeled it with a piece of tape that says (to my kids), "Remove this from the Birdmobile and YOU shall feel my wrath on a cellular level."

BIRDING BUDDIES: Go birding with friends, family, loved ones. It's more fun. You'll learn more stuff. And they might bring better snacks.

FIRST-AID KIT: This might not be obvious to take along on a birding trip, but I can tell you that there are times when a first-aid kit has been a life saver for me and for friends of mine. A basic kit should include adhesive bandages, antiseptics, joint wraps, tweezers, scissors, matches, a space blanket, flashlight, emergency phone numbers, a food bar or two, and aspirin or ibuprofen. I also always have jumper cables (for dead car batteries), a road flare, a pocketknife, and a length of nylon rope. And while you're at it, throw in a couple of quarters just in case.

IDENTIFICATION BASICS

Identifying birds is at the very heart of bird watching. Each bird encountered is like a little puzzle or mystery to solve, because, while birds of a single species all share a certain set of physical traits, no two individual birds, like no two individual humans, are exactly alike. You solve the mystery of a bird's identity by gathering clues, just like a detective.

Most of the clues we birders use are called *field marks*. Field marks are most often physical things we can see—visual clues such as a head crest, white bars on the wing (called wing bars), a forked tail, patches of color, spots on a breast, rings around eyes (yes, called eye-rings), long legs, and a curved bill, for example. But field marks also include behavior, such as hovering in flight, probing with a bill, pecking on a tree (woodpecker!), and flitting about actively. And field marks include sounds, too—songs, calls given in flight, chip notes, and even the whistle of wings. When added up, these clues should lead you to a correct answer: the bird's identity.

New birders are wise to start with the obvious visual field marks of a bird. You'll want to collect these field-mark clues in a logical way: Start at the top of the bird (by the head and bill) and work your way down and back. Most North American birds can be identified by field marks above the bird's shoulders: on the bill, head, and neck, and near the bend of the wing.

Tip: Resist the urge to take a quick glance at a strange bird, note one field mark, then drop your binocs and grab the field guide. Birds can fly and are prone to sudden movements. Field guides are books and cannot fly. Watch the bird as long as it lets you, *and then* reach for the field guide.

Some birds may appear to be completely plain and will require a longer look. But plainness itself can be a field mark! Don't give up. The clues are there, waiting for you to notice them.

Remember: Look carefully at your mystery bird, gather the clues, and then refer to your guide to solve the mystery.

PARTS OF A BIRD

Bird watchers who know the basic parts of a bird will find it easier to describe and identify unfamiliar birds. The terms used to label this Chipping Sparrow are commonly used in this book, in most field guides, and among bird watchers in the field.

The challenge of identifying birds is one of the best parts of bird watching. It can seem difficult and frustrating, but there are a few hints that will make it easier and more enjoyable.

The steps for identifying a bird are the same no matter where you are, no matter what bird you are watching. Here are some basic steps to follow.

SIZE: The first thing to notice is the size of the bird. Size will narrow down your choices a lot. To start with, think of birds as falling into the categories of small, medium, and large. Try associating those categories with objects you are familiar with: a cell phone, a soda can, a loaf of bread. It won't be long before your judgment of a bird's size is automatic. In most field guides, the size given is for the bird's length, measured from the tip of its bill to the end of its tail.

Here is an important point about judging size: A bird hunched over on the ground picking up seeds appears shorter and fatter than the same bird perched on a tree limb. Birds startled by a sudden noise or the appearance of a predator will stretch their necks, making them look considerably longer than when they are relaxed. Also, birds look thin when they are holding their feathers close to their bodies, and fat and round when they have fluffed them out, as they do in very cold weather. The key to judging size is to watch the bird for several minutes and to look at both its length and its bulk.

LOOK AT THE BIRD: Start with a general impression: What is *the* most noticeable thing about this bird? The answer to this question is a basic description of the bird's shape, size, and appearance. For example: This is a big, tall, thin bird with long legs.

Sometimes the general impression is enough; it is certainly a good starting point. Begin at the front (the head or bill) of the bird and work backward. The key to identification is often in the pattern of the head. Does the bird have stripes on the head? A line over the eye? Is there a noticeable color on the face or head?

Pay particular attention to the bill. Bill shape and size often indicate the family to which a bird belongs. A family is made up of bird species that are closely related and that share many characteristics. For example: All sparrows have short, thick bills. Warblers have short, thin bills. Thrashers and mockingbirds have long, thin bills, usually down-curved.

After you have looked at the head, check the back, wings, and underparts. Ask yourself: Is the back darker or lighter than the head or the belly? Does it have streaks? Are the underparts plain, or are there streaks or spots?

Wings often provide a key identification feature: the presence or absence of wing bars. Wing bars are contrasting, usually pale, lines across the wings. Many groups of birds—warblers and sparrows, for example—are divided, for ID purposes, into those that have wing bars and those that do not.

Last, look at the tail. Is it long or short, rounded or forked, darker or lighter than the back? Are there white panels or spots? Is it all one color? Does the bird bob or wag its tail persistently? Is the tail held cocked up or angled down?

All this sounds like a lot to remember, but after a few tries these questions will become automatic. As in most things, there is no substitute for practice. Try these steps several times on familiar birds. Remember, the key is to look for the *most obvious clues* to a bird's identification.

RANGE: Expect the expected. A bird's range (where it is regularly found) can be a valuable clue to its identity. If you live in Florida and identify a bird at your feeder that, according to the field guide map, occurs only in Oregon, your identification may be incorrect. Reconsider the other clues you have and try again. Birds occasionally show up a long way from where they are supposed to be, but it isn't common.

LOOK AT THE BOOK: Now it is time to use your field guide to identify the bird you have been watching. At first, the number of choices can be confusing. For new bird watchers, one of the best ways to find a species in a field guide is simply to start at the beginning and work through to the end. Don't worry; it won't be long before you are instinctively turning to certain sections. Small brown birds with thick bills will have you checking the sparrows, and chunky gray-on-brown birds with long tails and bobbing heads will have you focusing on the pigeons and doves.

Three Tips for Using Your Field Guide

1. The first several times you consult your field guide, be sure to go all the way through, even if you think you have found the bird. Also check to see if there are similar species with which the bird might be confused. A common mistake many people make is to settle on the first bird in the book that looks something like the bird they have seen.

2. The bird may not be identical to the picture in the book. Birds, like people, are variable. If almost everything matches and there are no very similar species, then you have almost certainly got the right bird.

3. If the possibilities have been narrowed down to three or four birds but you're still not sure, check another book. Every field guide has information that others lack.

Misidentifying Birds

Everyone does it. Beginners do it frequently, but even experts make mistakes. Misidentifying birds is part of the learning process. Don't worry about it. The occasional mistake will make you a better birder in the long run.

Getting better at bird watching is simply a matter of practicing as much as possible. Once you get the hang of it, you may find, as I do, that you're never *not* watching birds (and this can cause a slight problem if it's done during work, while driving your car, in the middle of an important conversation, or at other inappropriate times). The point is to look at birds every chance you get.

Finding the Bird in Your Binocs

Practicing with your binoculars is very important. Many new birders get frustrated at trying to find birds through their binoculars. They see a bird, quickly raise the optics to their eyes, and scan in crazy circles trying in vain to find the bird. Here's how to avoid this common problem: Lock your eyes on the bird, as if you are engaging in a staring contest. Don't move your eyes, your neck, or your head. Then slowly bring the binoculars up until they are aligned with your eyes. Did it work? If not, try practicing this same technique on a stationary object, such as a building or a tree. Trust me, this motion becomes automatic very quickly.

As you become more adept at finding birds in your binoculars, you may want to try this trick: When you see a bird at a distance, say, flying into the top of a distant tree, note where the bird is in relation to another noticeable object or feature, such as a clump of dead leaves, a notch in the tree's shape, or the tip of a passing cloud. These visual reference points will help you adjust your aim to find the bird more quickly. They also serve as excellent reference points to guide your fellow bird watchers to the bird, too.

Setting Up Your Binoculars

I spent my first five years of bird watching using my binoculars the wrong way. No, I wasn't looking in the wrong end! I had not adjusted

Binoculars magnify our view of birds so we can see them better.

my binoculars to fit my eyes. This resulted in a half decade's worth of headaches and lots of missed birds. After a fellow birder looked through my binocs and asked me how they were working for me, he showed me how to adjust them, and the relief was incredible! No more eyestrain or headaches and it was a *lot* easier to see the birds well and clearly.

Getting the most out of your birding binoculars requires a few minor adjustments the first time you use them. Start by adjusting the distance between the two barrels of the binoculars so that it is the right width for your eyes. Too far apart, and you won't be able to see through both barrels, but you will see large black areas on the outer edges of your view. Too close together, and you will see black edges in the center of your field of view. Forget those movie and TV scenes in which someone looks through binoculars and the camera shows their view as two partial circles surrounded by black. If you have the spacing right, *your view should be a single perfect circle* with no black edges in the center and an even black edge around the outside.

FOCUS: Next, find something to focus on, such as a dark tree branch against the blue sky, a street sign, or an overhead wire. Focus, with both eyes open, by turning the central focus wheel.

Now it's time for the fine focusing. First, close your right eye, and then, using only your left eye, adjust the focus wheel until your chosen object is in focus. For some birders it's more comfortable to hold a hand across one barrel rather than closing one eye. Next, you'll use the diopter—a secondary focus wheel, usually located on the right eyepiece. (Note: some models place the diopter on the central stem of the binoculars, or as a separate dialing feature on the central focus wheel.) The diopter compensates for the strength difference between your two eyes, giving you a finely focused image. Now, close your left eye, open your right eye, and adjust the diopter to bring your view into sharp focus.

Now open both eyes and see if your focus is crystal clear. If the image is not clearly focused, repeat these steps, making small adjustments using one focus wheel or the other. Once you have your binoculars focused properly, you'll use only the central focus wheel. The diopter setting will stay put.

How can you tell if your focus is correct? First of all, with the diopter adjusted, the view through your binoculars should appear almost three-dimensional. It should really pop with clarity. Your eyes should not have to work very hard when using your binoculars. If your eyes feel tired after a bit of birding with your binoculars, check the focus again. If the problem does not go away, you may need to have a professional check your binoculars for alignment.

Tip: If you see black every time you raise the binoculars to your eyes, your eyes may be too close to the lens. Roll or twist out the eyecups until you stop seeing black. (If you wear glasses, set the eyecups flat against the binoculars.)

Other Optics Tips

Clean binoculars make for happy birding. Protect your optics at all times, and clean them frequently. Use a soft brush (to get rid of dust and dirt), and then clean the surfaces with a soft cloth and lens fluid (available at most pharmacies or from your eye doctor).

Always wear your binoculars with a neck strap or a binocular harness. This is the safest and most convenient way to carry your binocs when birding. Don't carry them by the strap—that just puts them in danger of banging on things or being dropped. And don't carry them in your pocket, because you won't be able to get them up to your eyes fast enough.

A spotting scope is a fantastic tool for watching distant birds, such as ducks and shorebirds, and for scanning large expanses of habitat. You can certainly watch birds without a scope, and the cost and effort involved in using one can be considerable. Before you buy a spotting scope, try out several models owned by birding friends and see which one you prefer.

Get Field-Guide Familiar

Take some time when you're not bird watching to look through your field guide and get familiar with how it is organized. Most guides are organized taxonomically, with closely related bird species grouped together. But this won't seem logical until you are familiar with the taxonomic order of birds. Getting to know where the hawks, ducks, warblers, and sparrows are in your field guide will be very helpful to you when you need to find a bird quickly. Besides, it's fun to flip through a field guide and dream about seeing some of the amazing birds it contains.

Here's a tip that veteran birders often forget: *Use the guide's index!* Some guides have color-coded page edges to help you find each bird family. Others just have an index at the back, telling you the page number for each entry. Flipping through the pages to find your bird may seem smart, but I guarantee that using the index is faster.

Join a Club and Go Birding a Lot

Getting out in the field with more experienced bird watchers is the fastest way to improve your skills. Don't be afraid to ask questions ("How did you know that was an Indigo Bunting? How could you tell that tiny speck in the sky was a Red-tailed Hawk?"). Don't worry if you feel overwhelmed by the amount of new information. All new bird watchers experience this. When it happens, relax and take some time simply to watch the birds and your fellow birders.

I joined a bird club when I was 10 years old, and I'm still a member today. Most of my best field trips and some of the most memorable birds I've ever seen came to me because I was part of an active bird club. Find a club near you (most clubs have a website) and get out on a club field trip. (Try using Google to search for the name of the nearest bird club.) Introduce yourself as a new bird watcher, and, because birders are the friendliest folks on the planet, you'll soon have a batch of new birding pals eager to help you see more birds and have more fun.

Notes and Memory Devices

All bird watchers can benefit from taking notes about birds. These may be simple ID tips added to your field guide or field sketches and longer notes

Joining a bird club for field trips is both fun and helpful.

in a separate notebook or field journal. The point is to help you remember what you've seen and learned.

Another way to remember is to use memory devices. My birding mentor, Pat Murphy, had a memory device for almost all the birds she taught me. I learned to remember the difference between Downy and Hairy Woodpeckers as this: *Downy is dinky, Hairy is huge*. It also works for bird songs. *Pleased to meetcha, Miss Beecher* is how I remember the song of the Chestnut-sided Warbler, and *who cooks for you-all!* is the call of the Barred Owl. You may prefer to come up with your own memory devices. But whatever you do, write them down so you won't forget.

Don't Forget to Listen

When you're out watching birds, don't forget to use your ears, too. You'll frequently hear birds before you see them. (And sometimes you'll hear them and never see them.) Even if you don't immediately recognize the song or call, you can track down the bird by its sound and then let your eyes take over. As your birding enjoyment grows, you'll probably want to take advantage of the many sound recordings of birds that are available. Your school or public library will have some bird-sounds CDs, and there are dozens of others available at nature centers, at wild-bird stores, and on the Internet. I find my ears are better prepared for the sounds of spring migration when I've spent a bit of time in early spring listening to (and relearning) the songs I have on CD and on my smartphone.

Dress Right

Nothing can spoil a fine day of birding like being underdressed for cold or wet weather. Veteran birders know to dress in layers so they can adjust their outerwear to fit the outside temperatures. But you can't put on more clothes if you don't have them with you! Wear comfortable, supportive shoes and all the appropriate outerwear when you head out to see birds, and you'll avoid the pitfall of being underdressed. It's hard to hold binoculars steady when you're shivering!

Take It Easy

Finally, when you're bird watching, have a relaxed attitude and expect to have a good time. Don't get upset if you can't identify every single bird you see. Nobody can! I know lots of supercompetitive birders who have a hard time having fun when the birding is slow or when they miss out on a good bird. Sometimes, the harder you try, the harder it is to have fun. Remember, it's not always about seeing the most birds; it's about enjoying the birds you see.

BIRDING MANNERS

Out in the field with other bird watchers, it's a good idea to remember a few basic manners. These are mostly commonsense rules, but they're helpful to know.

KEEP YOUR VOICE DOWN. Seems basic enough, but excitement at seeing a new bird can sometimes make us exclaim, "*Hey! Look at that Bald Eagle!*" And this makes all the other birds fly away.

TREAT OTHERS AS YOU'D LIKE TO BE TREATED. If a fellow birder makes a bad call (and we all do this from time to time), don't tease him about it. This is the Golden Rule of Birding. I have personally seen some of the world's greatest bird watchers make bad calls—it happens. And when it happens to you, your fellow birders will treat you as you have treated them.

STAY WITH THE GROUP. If you are birding with a group of bird watchers and you wander ahead of them, you may see some birds they don't, but you may also scare off those same birds, which will upset your pals who didn't get to see them.

SHARE THE SCOPE. Because not every bird watcher has his or her own spotting scope, there is normally some sharing of the scope when a cooperative bird is found. When it's your turn at the scope, get a good look, but don't hog it. Once everyone else has gotten a look, you can take another turn.

HELP BEGINNERS. If you stick with birding long enough, you'll soon be in a position to help new bird watchers. Most bird watchers were helped out by a birding mentor or by someone more experienced when we started watching birds. When it's your turn to be the mentor, repay the favor.

PISH IN MODERATION. *Pish* is the sound we birders make to lure curious birds into view. To pish, softly say the word *spish* or *pish* through your clenched teeth. Many songbirds hear this sound and think it's a chickadee scolding a predator, so they pop up for a look-see. Some bird watchers prefer to make a soft squeaking noise instead. This can also work. Still others prefer to use a recording of bird songs or sounds to lure birds into view. In recent years, the iPod, smartphones, and other digital sound devices have become popular with birders because they can play hundreds of bird songs at the touch of a button. All of these methods of bird luring can be effective, but we need to use them in moderation. Overplaying a song can wear out a territorial warbler or other songbird, who will try to locate the invisible rival singing in his territory. We owe it to the birds we love so much to respect their privacy. So please pish or play songs in moderation. Your fellow birders can help you know when to say when.

BIRDING BY HABITAT

All birds have specific habitat preferences. Some are specialists: A large woodpecker, such as a Pileated, needs old woods with big trees. Others are generalists: Song Sparrows may be found in city parks, suburban gardens, and brushy field edges in farmland.

When you're in the field bird watching, a great way to prepare yourself for the species you may encounter is to think like a bird. Look at the landscape and all the various habitats it contains and think about where you would go if you were a certain bird. Combine this with your previous experiences encountering bird species in particular habitats, and you are officially "birding by habitat."

Birds' habitat preferences can change seasonally. For example, Eastern Bluebirds prefer open grassy fields and meadows in spring and summer, but in late fall and winter, they move to wooded habitat for shelter and to find food. Many of the migratory birds, including many of our most colorful songbirds, spend spring and summer with us, then leave our midst altogether to spend the winter in the warm, insect-rich tropics.

The illustration on the following pages shows some of the birds to expect on a summer day birding along a country road in the eastern half of North America. These same concepts work elsewhere around the continent; while the habitats may look a bit different, the bird species that prefer these specific habitats will remain consistent. Try birding by habitat the next time you are out bird watching. It's easy to do, and you can't help but become a better birder by doing it.

1
2
3

15
11
12
14
10
13
4
6
5
7
9
8

KEY TO ILLUSTRATION

1. SYCAMORE TREE
- Baltimore and Orchard Orioles
- Yellow-throated Warbler, Northern Parula
- Wood Duck
- Eastern Screech-Owl (in natural cavities)

2. PINES AND CONIFERS
- Owl roosts, nests
- Heron nests
- Raptor nests
- Pine Warbler, Black-throated Green Warbler
- Chickadees and nuthatches

3. BEAVER SWAMP
- Swallow nests
- Ducks
- Spotted or Solitary Sandpiper
- Woodpeckers (on dead trees)
- Herons
- Flycatchers

4. BEAVER POND
- Ducks
- Swallows
- Herons
- Grebes
- Shorebirds

5. CATTAIL MARSH
- Rails
- Blackbirds
- Marsh Wren
- Herons, bitterns

6. TUSSOCK (SEDGE) MARSH
- Swamp Sparrow
- Sedge Wren

7. CUT BANK OF STREAM
- Nesting Belted Kingfisher
- Nesting Bank Swallow
- Rough-winged Swallow
- Spotted Sandpiper

8. GRAVEL ROAD/ROADSIDE
- Eastern Bluebird
- American Goldfinch
- Killdeer
- Ruby-throated Hummingbird
- Indigo Bunting
- Common Yellowthroat

9. BRUSHY OLD FIELD
- Common Yellowthroat
- Prairie Warbler
- Blue-winged Warbler
- Song Sparrow
- Brown Thrasher
- Yellow-breasted Chat
- Northern Cardinal

10. FARMYARD
- Barn Swallow nests
- Cliff Swallow nests
- Barn Owl nests
- American Kestrel nests
- House Sparrow
- European Starling
- Brown-headed Cowbird
- Rock Pigeon

11. HARDWOOD FOREST

- Scarlet and Summer Tanagers
- Warblers
- Vireos
- Thrushes
- Flycatchers
- Cuckoos
- Whip-poor-will, Chuck-will's-widow
- Woodpeckers
- Chickadees
- Nuthatches
- Titmice

12. HAY MEADOW

- Eastern Meadowlark
- Dickcissel
- Blackbirds
- Field Sparrow
- Grasshopper Sparrow
- Eastern Bluebird
- Eastern Kingbird
- Bobolink

13. POWER LINES

- Hawks (perching)
- Eastern Bluebird
- Indigo Bunting
- Eastern Meadowlark
- Mourning Dove
- Eastern Kingbird
- Blackbirds

14. PASTURE

- Killdeer
- Eastern Meadowlark
- Brown-headed Cowbird
- European Starling
- Eastern Bluebird
- Prairie Warbler
- Field Sparrow

15. SKY

- Soaring raptors (hawks and vultures)
- Swallows

FIVE OUTSIDE-THE-BOX TIPS FOR IMPROVING YOUR BIRDING SKILLS

1. GO WITHOUT. Now that I've spent the past 25 pages sermonizing about always having your binoculars and field guide with you, I'm going to encourage you to go outside without them. It's funny but with our powerful optics magnifying our views of birds, we sometimes miss the beauty of the birds in their natural context. My favorite all-time view of a Brown Thrasher was on a bleary, rainy April morning when a male, just back from his winter home down south, began his spring song from the sumac-covered edge of our meadow. I was outside without binoculars, working in the garden. In my memory, that will always be the quintessential Brown Thrasher moment. And I never got binocs on him.

2. GET A SECOND PAIR OF BINOCULARS FOR YOUR CAR, OFFICE, KITCHEN WINDOWSILL, BRIEFCASE, ETC. Always having optics handy means you'll miss fewer birds. My car binocs have netted me several wonderful birds, including flyover tundra swans in an ice fog, a Golden Eagle surprise along a highway, and a flock of black terns migrating up the Ohio River as I was heading out to a business lunch.

3. SET GOALS; CHALLENGE YOURSELF. A friend of mine had a banner year in our southeastern Ohio county last year. During the course of trying to see as many birds in our county in one year as he possibly could, he learned a lot about the seasonal abundance of birds, discovered some new local birding hotspots, and made several new birding friends. Try setting your own goal: to see all the woodpeckers in your area; to add five new species to your yard list; and so on. Just make sure you're having fun while doing it.

4. GET OUT OF HERE! Consider taking your bird watching away from your usual haunts. Visit a different habitat in your county. Visit a refuge or wildlife area you've never seen. Better yet: go to a birding festival! There are more than 300 birding and nature festivals held annually in the United States. Most offer guided field trips and entertaining and informative presentations. Some have vendors selling their products and services. Not only will you expand your birding horizons, you'll probably also make some new birding friends along the way.

5. GET CREATIVE! Add a bit of creativity to your bird watching by writing about, drawing, or photographing birds. To know a bird fully, you must look at it with new eyes. What better way than by trying to draw it, photograph it, or write about it? I have friends who write daily haikus about birds. I know others who share daily bird photos online. I—and many of my birdy friends—post regularly on Facebook, Twitter, and our blogs about our birding experiences, favorite birds, and birding places. For examples on how to do this, just get online and do some basic searches for birding content. Being creative with birding seems to add a wonderful dimension to an already rich, engaging hobby.

BE GREEN: TEN THINGS YOU CAN DO FOR BIRDS

1. CREATE BIRD-FRIENDLY HABITAT. There are countless ways to create habitat for birds in your backyard. Perhaps the easiest is to let things go wild in one part of your property. Chances are the plants that grow in your wild area will be natural sources of food for the birds. A more focused approach involves providing birds with the four things they need: food, water, shelter, and a place to nest.

2. NO CHEMICALS! It's widely known that seemingly safe lawn and garden insecticides and herbicides can be harmful to birds. Many of these chemicals target the pests that are food sources for birds, so any birds eating treated insects or seeds are also ingesting toxic chemicals. Avoid or at least minimize the use of toxic lawn and garden chemicals.

3. RECYCLE YOUR TRASH. Each plastic, glass, aluminum, or tin item you recycle is one less piece of trash cluttering up the planet and one less ugly and hazardous item that we (and the birds) have to deal with in the environment. Recycling also saves money, eases pressure on habitat, and reduces pollution created by the production of first-generation materials such as glass, tin, plastic, and aluminum.

4. KEEP YOUR FEEDERS AND NEST BOXES CLEAN. A once-a-month scrub cleaning of bird feeders will go a long way toward reducing disease transmission. Use a solution of one part bleach to nine parts hot water. Keeping your nest boxes clean is equally important. Clean out old nesting material several weeks after the nesting season is over. If the inside is really fouled with droppings, clean it out with the same bleach solution described above. Replace the old nesting material with a fresh handful of dried grasses to give the birds some insulation if they use the box for fall and winter roosting.

But how do you know when the nesting season is over? Read on . . .

5. MONITOR YOUR NEST BOXES. Cavity-nesting birds face almost constant competition from non-native species that want to use these same cavities (hollow trees, old woodpecker holes, and nest boxes) for nesting. By checking your nest boxes regularly, you can discourage these introduced species and keep your nest boxes available for native species that need a place to nest or roost. Chickadees, titmice, nuthatches, woodpeckers, Tree and Violet-green Swallows, Purple Martins, wrens, and bluebirds are among the species that commonly use backyard nest boxes.

Monitoring your nest boxes regularly is good for the birds (and for you, too).

6. PARTICIPATE IN BIRD COUNTS. There are dozens of local, national, and even international bird counts in which bird watchers can play a part. The National Audubon Society's annual Christmas Bird Count is one of the longest-running counts. The Cornell Laboratory of Ornithology conducts Project FeederWatch and the Great Backyard Bird Count as well as several other specific annual counts. See Resources, on page 348, for contact information.

7. REDUCE WINDOW KILLS. Mylar strips, crop netting, branches, screens, and hawk silhouettes have been suggested as foils to keep birds from flying headlong into your windows. Placing these items outside, in front of the problem panes, breaks up the windows' reflections of the surrounding habitat so that the windows do not fool birds into flying into them.

8. KEEP CATS INDOORS. Even the most slothful, couch-potato cats can catch birds if given a chance. It's been estimated that housecats kill many millions of birds each year—deaths that could be avoided if these pets were kept indoors. For more information, write to Cats Indoors! Campaign, American Bird Conservancy, PO Box 249, 4249 Loudoun Avenue, The Plains, VA 20198-2237; or go to www.abcbirds.org/cats.

9. SUPPORT CONSERVATION INITIATIVES. Every day there are a thousand battles we bird watchers can fight on behalf of birds. The key is picking your spots so that you can make the most effective impact. Not all conservation initiatives are created equal, so be sure you're fully informed about the issues. In most cases, if bird habitat is preserved or created, it's a good thing. After you've created healthy habitat for birds in your own backyard, you may wish to contact the American Bird Conservancy or The Nature Conservancy to see how else you can help.

10. MAKE A NEW BIRD WATCHER TODAY. Why not take a friend along on your next bird-watching trip, to the next bird-club meeting, or on a tour of your bird-friendly backyard? The more bird watchers we have today, the more good we can do for the birds tomorrow. It is especially important to encourage young people to get engaged with the natural world. I've spoken to a lot of bird clubs and birding festivals, and let me tell you, the number of older birders vastly outnumbers the younger attendees. There are young-birder clubs all over North America. Please consider supporting their efforts to cultivate an interest in birds among young people.

HOW TO USE THIS GUIDE

This is not the only field guide you will ever need. In fact, I hope it's the first of many bird books you'll want to have. *The New Birder's Guide* is aimed at giving beginning bird watchers an easy first step into the world of birds.

The 300 species contained in these pages were chosen because they are birds that either are commonly seen or that every new birder should get to see in his or her lifetime. They are organized in general taxonomic order and follow the species names and Latin names currently agreed upon by most ornithologists.

In order to make this book small enough to be easily carried and used, I've limited it to slightly more than 300 species. I guarantee that while out birding, you will encounter birds that are not in this book. This is the first clue that you need to get a more comprehensive field guide covering *all* of the birds of your region, or even *all* of the birds of North America.

Each bird's page has these main parts:

- A species profile
- The species' common name and Latin name, along with body length for size reference
- One or two photo images of the typical plumages
- One black-and-white drawing of the bird doing something interesting
- Range map showing seasonal distribution

And each species profile has these sections:

- Look for (field marks for identification)
- Listen for (the bird's song and other sounds)
- Find it (its habitat preferences and seasonal occurrence)
- Remember (an additional ID tip, often comparing it to similar species)
- Wow! (an interesting extra fact about this bird)

THE BEST BIRDING TIPS

Keeping Your Bird List

Most bird watchers enjoy keeping a list of their sightings. This can take the form of a written list, notations inside your field guide next to each species' account, or a special journal meant for just such a purpose. There are even software programs available to help you keep your list on your computer. In birding, the most common list is the life list. A life list is a list of all the birds you've seen at least once in your life. Let's say you noticed a bright black and orange bird in your backyard willow tree one morning and then used your field guide to identify it as a male Baltimore Oriole. If it's a species you'd never seen before, now you can put it on your life list. List keeping can be done at any level of involvement, so keep the list or lists that you enjoy. Many birders keep lists for fun—all the birds seen in their yard in one day, for example. Others keep lists for their county, their state or province, or for a complete year.

At the bottom of each bird's page in *The New Birder's Guide,* you'll find a check box and space for a short note. You can use this to start your life list.

A big part of the fun of birding is birding with friends.

To get the most out of this book, take it with you whenever you go birding. Flip through it when you are at home or when you can't be outside watching birds. Though we're all told never to write in a book, please write notes in this one. Record your bird sightings at the bottom of each bird's page. Make it your own book. If we've done our work properly, you'll soon progress beyond *The New Birder's Guide* and will move on to larger, more complete field guides covering all the birds of North America.

TEN TIPS FOR BEGINNING BIRD WATCHERS

1. Get a decent pair of binoculars, one that is easy for you to use and hold steady.

2. Find a field guide to the birds of your region. (Many guides cover only eastern or western North America.) Guides that cover all the birds of North America contain many species that are uncommon or entirely absent from your area. You can always upgrade to a continent-wide guide later.

3. Set up a basic feeding station in your yard or garden.

4. Start with your backyard birds. They are easiest to see, and you can become familiar with them fairly quickly.

5. Practice your identification skills. Starting with a common bird species, note the most obvious visual features of the bird (color, size, shape, patterns in the plumage). These features are known as field marks and will be helpful clues to the bird's identity.

6. Notice the bird's behavior. Many birds can be identified by their behavior—woodpeckers peck on wood, kingfishers hunt for small fish, swallows are known for their graceful flight.

7. Listen to the bird's sounds. Bird song is a vital component to birding. Learning bird songs and sounds takes a bit of practice, but many birds make it pretty easy for us. For example, chickadees and Whip-poor-wills (among others) call out their names. The Resources section of this book contains a list of tools to help you learn bird songs.

8. Look at the bird, not at the book. When you see an unfamiliar bird, avoid the temptation to glance at the bird and then grab the guide. Instead, watch the bird carefully for as long as it is present. Your field guide will be with you long after the bird has gone, so take advantage of every moment to watch an unfamiliar bird while you can.

9. Take notes. No one can be expected to remember every field mark and description of a bird. But you can help your memory and accelerate your learning by taking notes on the birds you see. These notes can be written in a small pocket notebook or even in the margins of your field guide.

10. Venture beyond the backyard and find other bird watchers in your area. The bird watching you'll experience beyond your backyard will be enriching, especially if it leads not only to new birds but also to new birding friends. Ask a local nature center or wildlife refuge about bird clubs in your region. Your state ornithological organization or natural resources division may also be helpful. Being with others who share your interest in bird watching can greatly enhance your enjoyment of this wonderful hobby.

DEBUNKING SOME BIRD MYTHS

ALL BIRDS SOUND ALIKE. This may seem true to the ears of a relatively new birder, just as a foreign language can sound like gibberish. Once you have some experience in listening, you'll find there are vast differences in the way birds sound—even closely related species. The key is to use your ears when bird watching as often as you do your eyes. For me all warbler songs sounded alike until one day when something clicked and they no longer did. It took a few springs of confusion and practice, but eventually I could identify birds by sound.

I CAN'T IDENTIFY SPARROWS (OR FALL WARBLERS, OR SHOREBIRDS, OR GULLS)! As with the bird sounds, the visual clues to identification range from obvious to subtle. All sparrows may seem like Little Brown Jobs (or LBJs), but if you choose a common species and learn to recognize it reliably, you can use it as a reference point for noticing how other sparrows are different. For me the reference bird is the Song Sparrow. I know Song Sparrows very well, and this makes it easier for me to recognize a bird as *not* being a Song Sparrow. Then I set about discerning what the visual differences are, and soon I've narrowed the choices down to just one or two birds.

MY OLD BINOCS ARE JUST FINE; I DON'T NEED NEW ONES. If you are using binoculars that are more than 15 years old, you might want to consider an upgrade. In the past decade, the technology and engineering of optics manufacturing have created a surge in excellent, affordable binoculars. For $300 you can now get optics that are better than anything that was being sold 15 years ago. Go to an optics retailer, a sporting goods store, or a bird festival trade show and try out some of today's amazing binocs. You won't believe how much better the birds look.

Backyard Bird Myths

There are three enduring bird myths that are generally associated with backyard birds.

MYTH #1: **If you stop feeding the birds, they will starve.**

Actually, birds have evolved as highly mobile creatures, completely adept at flying from one food source to another. So it's okay to go on vacation. Your birds will come back once you return and refill the feeders.

MYTH #2: **If you don't take down your hummingbird feeders, the hummers won't migrate, and when it gets cold they will freeze to death.**

Nothing short of illness or injury will stop a bird from migrating if it feels the natural urge to go. Hummingbirds are no different. They have internal clocks, triggered by the changing levels of daylight, which tell them when to migrate. Your feeder holds no sway over them. And when they *do* migrate they do so under their own power. They do not migrate on the backs of Canada Geese.

MYTH #3: **Don't feed peanut butter to birds. It sticks to the roof of their bills and they could choke.**

Not true, though regular roasted peanuts out of the shell are a cheaper (for you) and easier (for the birds) way to offer the fat and energy of the peanut to a wild bird.

- You recall the birds on the list you kept during your last family reunion more vividly than the name of your great-aunt from California.

- You realize that you *might* remember her name if she had California Quail in her backyard.

- You tell your doctor (without any hint of humor or irony) that you're suffering from warbler neck.

- You have to explain to the highway patrolman that you weren't texting while driving . . . you were birding.

- Every coworker asks you a bird question within five minutes of starting a conversation.

- Your dog is named Towhee. Your daughter is named Phoebe. Your son is named Pewee.

- You've planned family vacations to fill in the holes on your life list.

- On your last tax return there were expense listings for "spotting scope," "bulk bird seed purchases," and "birding festival travel costs."

- You know the state birds for each of the 50 states, but you can't name the capitals.

- You know how to spell *pyrrhuloxia, phainopepla,* and *ferruginous,* but you can never remember if it's "i before e, except after c," or the other way around.

This Carolina Chickadee calls out chick-a-dee-dee-dee.

- You have experienced Optics Envy, Migration Withdrawal, and a Winter Finch Irruption and firmly believe they should all be covered by your health insurance.

- You would rather be birding than doing almost anything else. Including reading this book.

GREATER WHITE-FRONTED GOOSE

Anser albifrons Length: 28"

LOOK FOR: Named for the white forehead patch on its face, the Greater White-fronted Goose is a large bird that looks slender in flight. Field marks that set it apart from our other geese include the boldly patterned black-and-white tail with a white tip, a speckled dark-and-light belly, bright orange legs, and a pale pink bill.

LISTEN FOR: White-fronts make a high-pitched series of rapid, tooting cackles, nothing like the honking of a Canada Goose. Flocks are very vocal.

WOW!

The splotchy black, gray, and white pattern of this bird's belly has earned it the nickname Specklebelly.

REMEMBER: You might confuse this species with the similar Greylag Goose, a domesticated fowl. A single bird in a farmyard or city park is probably not a wild Greater White-fronted Goose.

◀ *Greater White-fronted Geese are often mistaken for the Greylag Goose (in back) and are often seen with the Canada Goose (right).*

FIND IT: A large population uses the Central Flyway to migrate between the Arctic tundra breeding grounds and the coastal Gulf of Mexico wintering grounds. White-fronts forage in large open fields and shallow marshes.

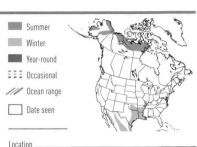

Summer

Winter

Year-round

Occasional

Ocean range

☐ Date seen

____

Location ______________

LOOK FOR: Named for its snow-white plumage, the Snow Goose comes in two colors (or morphs), white and dark, though white Snow Geese are far more common. The typical white morph has an all-white body with black wingtips, pink legs, and a large pink bill. The dark morph, or Blue Goose, has a dark gray body and white head.

◀ *For a long time, the Blue Goose (top) was thought to be a separate species from the Snow Goose.*

LISTEN FOR: A high-pitched *howk-howk* is the most common call, often heard from large flying flocks. Foraging birds utter piglike grunts.

REMEMBER: In the Midwest and West, the similar but smaller Ross's Goose may be found with Snow Geese. The Ross's has a smaller pink bill and looks like a miniature Snow Goose.

WOW!

Snow Geese have enjoyed a population boom in recent years, and this is putting a strain on the very sensitive tundra habitat where they breed.

FIND IT: Snow Geese nest on the tundra of the far north. Migrating flocks fly very high in long, wavering lines (rather than the V formations that some other geese prefer), which has earned this species the nickname Waveys.

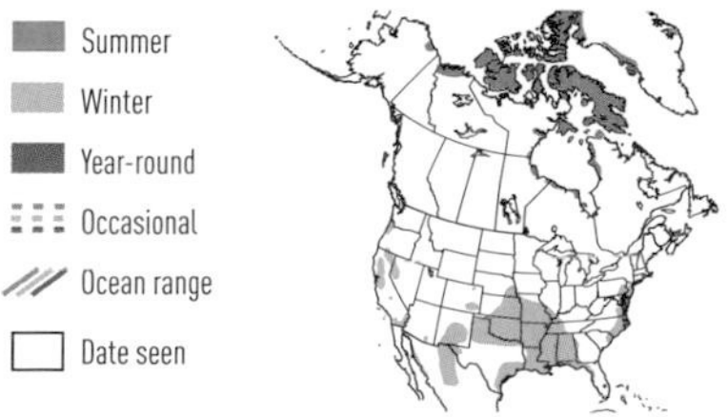

Date seen ____________

Location ____________

LOOK FOR: The Ross's Goose is like the Mini-Me of the Snow Goose, more petite in every way: smaller overall with a smaller, rounder head; shorter neck; and a shorter bill than the Snow Goose. The bill of the Ross's Goose is all pinkish, lacking the black "smile" patch of the Snow Goose.

LISTEN FOR: Nasal, high-pitched yelps. Not as vocal as the Snow Goose, but call is similar, pitched higher.

REMEMBER: When scanning through a large flock of Snow Geese, look for the smaller overall size and rounder shape of a Ross's Goose. The difference in head and bill size and shape is a good field mark to check on small white geese.

WOW!

Ross's Geese sometimes occur in the darker-bodied "blue" form, just as Snow Geese do. But this was not discovered until the 1970s!

► *A careful observer can pick out a Ross's Goose or two from most large flocks of Snow Geese.*

FIND IT: Often found in mixed flocks with Snow Geese in winter in fields, marshes, and bays. Nests in summer on the tundra in the far north. Ross's Geese are more commonly found in wintering flocks along the Mississippi Flyway and points west.

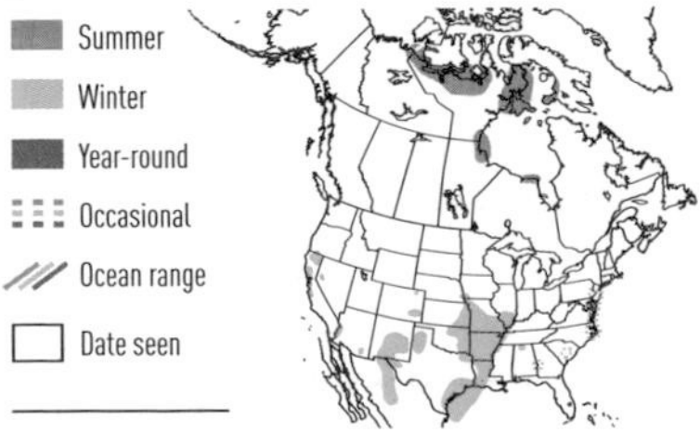

Location ______

LOOK FOR: Almost everything about this small, short-necked goose is dark: head, bill, body, wings, legs. The overall dark head and body contrast with a white lower belly and upper tail, giving flying birds a dark-in-front, white-in-back appearance. If you can get close to a Brant, look for a partial white necklace.

LISTEN FOR: A low, rolling honk that sounds almost like a goat or sheep.

REMEMBER: The Brant lacks the bright white cheek patch of the Canada Goose. Brant are also much smaller and shorter necked than Canada Geese.

WOW!

Migrating Brant may fly nonstop from their high-Arctic breeding grounds to the Atlantic Ocean, a distance of more than 1,800 miles.

▶ *Eelgrass is like spaghetti to Brant—they depend on it for much of their diet.*

FIND IT: Brant breed on the tundra. This small, dark goose is almost never seen away from coastal waters, where huge flocks of Brant can be seen flying in disorganized, dark, wavering lines. Foraging flocks use shallow bays and tidal mud flats.

Summer

Winter

Year-round

Occasional

Ocean range

Date seen

____

Location ______________

LOOK FOR: This familiar large goose, with its black head and neck and white "chinstrap," was once a symbol of wild North America. The dark gray-brown body contrasts with a white rump, the back third ending in a black tail. A black bill and black legs complete the Canada's typical plumage.

LISTEN FOR: A deep *honk* or *ha-ronk*, rising in tone, which is considered by many to be the classic goose sound. Flocks in flight are noisy. Flocks and families on the ground make a variety of grunts, gabbles, and hissing sounds.

REMEMBER: Flocks of Canada Geese flying overhead in a V formation may contain other species, including Snow Geese, White-fronted Geese, and other waterfowl.

WOW!

Canada Geese are so completely comfortable living near humans that they (and their droppings) have become a nuisance in many areas, including golf courses, airports, and city parks.

◀ *Canada Goose families often migrate together following the nesting season.*

FIND IT: Canada Geese occur from remote tundra and marshy regions to fields and wetlands. Canadas now maintain a robust year-round population across the middle third of the U.S. Some birds are nonmigratory.

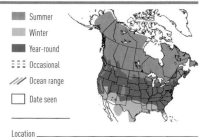

- Summer
- Winter
- Year-round
- Occasional
- Ocean range
- Date seen ____________

Location ____________________

MUTE SWAN

Cygnus olor Length: 60"

LOOK FOR: Our only large white swan with an orange bill, the Mute Swan was introduced to North America from Europe in the mid-1800s. The orange bill has a black knob on top, also unique. The Mute Swan swims with its neck in a graceful S-shaped curve, bill pointing down, and sometimes fluffs out its wing feathers, especially when confronting an intruder.

LISTEN FOR: Not mute, but does not possess a lovely song either. Grunts and hisses during aggressive encounters. In flight, wings make a low, humming sound.

REMEMBER: You are more likely to see Mute Swans floating on a quiet body of water rather than in flight.

WOW!

Though we think of a pair of Mute Swans on a park pond as a symbol of love, these birds aggressively drive other waterfowl away from their breeding territory. They are well-dressed bullies who do not play well with others.

◀ *Mute Swans harass native waterfowl, taking entire ponds as their exclusive turf.*

FIND IT: Found on quiet ponds, lakes, and bays, especially near human habitation. Equally at home on bodies of salt- or fresh water. Many birds are nonmigratory residents.

Summer

Winter

Year-round

Occasional

Ocean range

☐ Date seen

____

Location ______________

TUNDRA SWAN

Cygnus columbianus Length: 53"

LOOK FOR: The Tundra Swan is the smallest of our three white swans. The Tundra's black bill and shorter, straighter neck differentiate it from the orange-billed Mute Swan. The adult Tundra Swan has a small patch of yellow in front of the eye, but this can be difficult to see. Distinguish Tundra Swans from the much rarer Trumpeter Swan by the Trumpeter's much longer, tapered bill.

WOW!

If you live anywhere along the Tundra Swans' spring or fall migratory routes, you may hear the flocks of 100 birds or more calling as they migrate overhead.

LISTEN FOR: Tundra Swans give high-pitched, nasal *whoo-ooo* calls that almost sound like the trumpeting of a tiny elephant.

REMEMBER: Tundra Swans appear gooselike in flight, but the most similar goose, the Snow Goose, always shows clear black wingtips. Tundra Swans appear all white in flight.

▼ *Tundra Swans often time their migration just ahead of passing weather fronts.*

FIND IT: Tundra Swans prefer large bodies of water such as lakes, reservoirs, and bays. Nesting in the northernmost reaches of North America, they follow a variety of migration routes to three primary wintering areas.

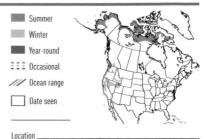

Summer
Winter
Year-round
Occasional
Ocean range
Date seen ________

Location ________

TRUMPETER SWAN

Cygnus buccinator Length: 60"

LOOK FOR: Though rare throughout their range, Trumpeter Swans are hard to miss. It is our largest native waterfowl species (the Mute Swan is as large but is an introduced species in North America). Its all-black bill is long and straight. Adults are all white; young birds show some brownish body feathers, then appear grayish before acquiring the all-white appearance of adults in their second summer.

WOW!

Hunting and market shooting nearly wiped out the Trumpeter Swan in the late nineteenth and early twentieth centuries.

LISTEN FOR: Call is a hornlike tooting, which gives the bird its name. Longer notes sound very much like a trumpet.

REMEMBER: Three things help to separate the Trumpeter Swan from the similar Tundra Swan: the Trumpeter is much larger overall; its bill is long, straight, and all black (Tundras often show some yellow in the bill); and the Trumpeter is much more rare and local than the Tundra.

◀ *Trumpeter Swan pairs stay together for as long as both partners live. Both parents share in the care of the young.*

FIND IT: Trumpeter Swans nest on wooded lakes and ponds in the western mountains and northwestern portions of North America, usually far from human disturbance. Some breeding populations have been reestablished in the Midwest through conservation-recovery programs.

Summer
Winter
Year-round
Occasional
Ocean range
☐ Date seen

____

Location ____________________

WOOD DUCK

Aix sponsa Length: 18"

Male (left), female (right)

LOOK FOR: A medium-sized duck that may hold the title of Most Beautiful among waterfowl, the male Wood Duck, in breeding plumage, is a rainbow of colors. In late summer, males in eclipse (that is, dull and drab) plumage resemble females. Wood Ducks are fast fliers and often call in flight.

WOW!

Just one day after they hatch, Wood Duck babies leap from the nest cavity.

LISTEN FOR: Female Wood Ducks give a high-pitched, excited, two-part whistle: *uh-wheek!* They also give a softer series of rapid whistles: *oh-oh-oh-oh.*

REMEMBER: In flight, Woodies appear long-tailed compared to other common ducks. Wood Ducks will use nest boxes, and human-supplied housing has contributed to this species' population comeback during the past 60 years.

▶ *Wood Duck ducklings leaping from a nest box. They are so lightweight that they float to a soft landing below.*

FIND IT: Considered one of our perching ducks for their habit of perching, roosting, and nesting in trees. A scan of the trees and snags in wooded swamps, quiet rivers, and in forested marshes will often reveal Wood Ducks dabbling or sitting quietly.

Summer

Winter

Year-round

Occasional

Ocean range

Date seen

____

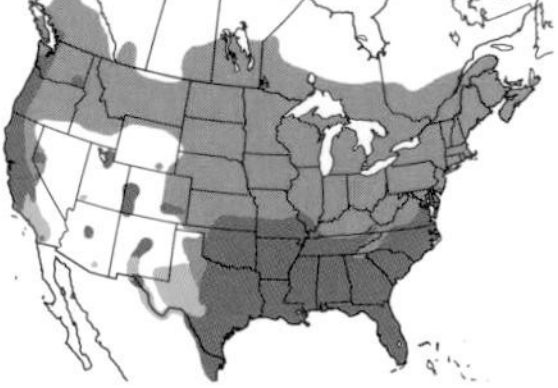

Location ______________________________

LOOK FOR: The white crown stripe of the male American Wigeon has earned this species the nickname Baldpate. This field mark contrasts with the dark eye patch and can be seen from great distances. Both male and female appear round headed and have a pinkish chest and body. The light blue-gray bill has a black tip.

LISTEN FOR: Females give a guttural growl. Males make a high whistle that sounds like a squeaky squeeze toy.

REMEMBER: In flight, drake (male) wigeons flash large white patches on the upper surface of their wings. Both sexes have underwings with white centers and both show white bellies.

WOW!

American Wigeons behave like bandits on large lakes. They are known to steal food from other waterfowl.

◀ *An American Wigeon steals a bit of food from a Canada Goose.*

FIND IT: Look for American Wigeons on large marshes, lakes, and ponds, where they forage near the surface. They also graze for food on land more commonly than other ducks.

Summer

Winter

Year-round

Occasional

Ocean range

Date seen

Location ____________

GADWALL

Anas strepera Length: 20"

LOOK FOR: At first glance, the Gadwall might look like nothing special. A closer look, especially at the male, or drake, reveals this duck's subtle beauty. The most reliable field mark is the male's black "underpants" (the back end, under the wings), visible from a great distance. The Gadwall's head appears lumpier than other ducks'. Females lack the black underpants but can be distinguished from other plain brown ducks by the light orange spots in the bill.

WOW!

What's a "Gadwall"? Nobody seems to know. The origin of this bird's name is a mystery.

LISTEN FOR: Courting males utter a low, nasal beep, often in a series. Females quack like Mallards.

REMEMBER: In flight, both sexes of Gadwall show white on the underside of the wings and a small white square on the upperwing, close to the body.

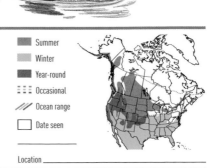

► *The Gadwall's white wing patch and the black underpants of the male can be seen from a great distance.*

FIND IT: Common on marshes and shallow lakes in open (not wooded) settings. Prefers fresh water. Most common from the Great Plains west, but many are found in the East during migration and across the lower third of the U.S. in winter.

Summer
Winter
Year-round
Occasional
Ocean range
Date seen ____________

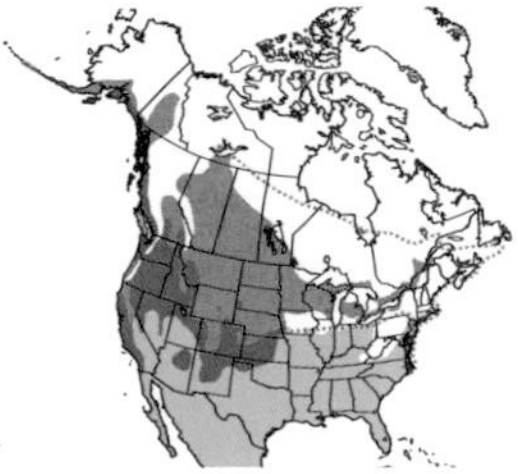

Location ____________________________

WOW!

Mallards and American Black Ducks often breed with each other, creating hybrid offspring. Over time, this interbreeding has caused an increase in Mallards and a decrease in Black Ducks.

LOOK FOR: The Black Duck is a medium-sized dabbling duck with a dark charcoal body. Males and females are very similar, though the male has a yellowish bill while the female's bill is dull gray. In flight, Black Ducks show mostly white underwings and mostly dark upperwings. This contrast gives a flashing appearance to their flight.

LISTEN FOR: A nasal, quacking call: *whap-whap, whap-whap.*

REMEMBER: Black Ducks and Mallards are closely related, and these two species are often found together. Telling a female Mallard from a female Black Duck is difficult: female Blacks are darker overall and have no orange on the bill.

◀ *The white wing linings showing on this flying American Black Duck are a key field mark for this species.*

FIND IT: Almost any body of water can host Black Ducks, but they seem to prefer coastal salt marshes and, inland, wooded wetlands. They are found year-round across the Midwest and along the Atlantic Coast from Maine to the Carolinas.

Summer
Winter
Year-round
Occasional
Ocean range
Date seen

____

Location ______________

MALLARD

Anas platyrhynchos Length: 23"

LOOK FOR: The male Mallard's green head and yellow bill make it one of North America's most recognizable bird species and certainly our most familiar duck. Domestic and semidomestic versions of Mallards exist in many farmyards, parks, and zoos, but few feature the clean-looking plumage of the wild Mallard.

LISTEN FOR: Female Mallards give the typical duck call: *quaaack!-quackquackquack!* This sounds very much like Donald Duck of cartoon fame. Males give a high-pitched *queeep!*

REMEMBER: Domestic Mallards can be found almost anywhere. Wild Mallards, with sharp-looking plumage, are wary birds that are quick to flush into the air.

▲ *A female Mallard leads her brood across a curb. The ducklings will follow their mother anywhere she goes.*

WOW!

Mallards are strong fliers and have been clocked at speeds of up to 60 mph!

FIND IT: Mallards can be found in almost any freshwater habitat. Some live on suburban ponds and nest in the shrubbery around buildings. Often seen in pairs, which form in fall or winter and last through the breeding season.

Summer

Winter

Year-round

Occasional

Ocean range

Date seen ________

Location ________

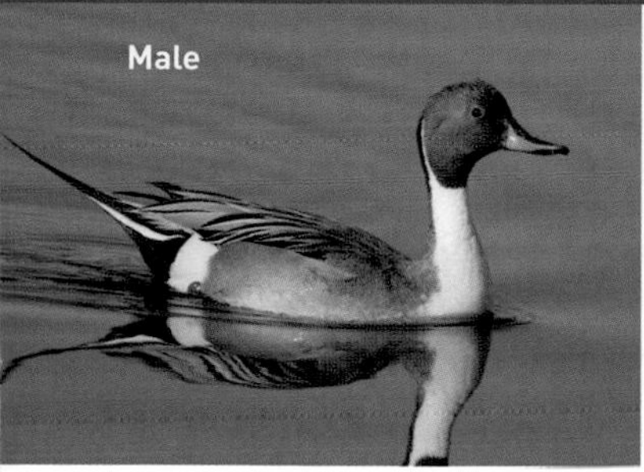

LOOK FOR: The long neck and long tail of the Pintail (especially of the male) give it an elegant shape that is distinct from other ducks. The male's chocolate brown head and white neck and breast are reliable as field marks in all seasons except late summer, when molting males resemble females. Female Pintails are toasted-marshmallow brown overall with dark gray bills.

WOW!

The male Northern Pintail's tail has two long central tail feathers that extend well past the other tail feathers. Other names for this species include Spring-tail, Spike-tail, and Sharp-tail Duck.

LISTEN FOR: Females give a typical ducklike *quack*. Males give a high-pitched *twoot-twoot* and a very unbirdlike rising and falling buzzy whistle.

▼ *A male Northern Pintail stretches its foot and wing past his long tail feathers.*

REMEMBER: Pintails are skittish—they are often the first birds to flush into flight from a pond or wetland containing lots of ducks.

FIND IT: Very adaptable in their choice of habitat, Northern Pintails can be found in marshes and farm fields, and on prairie ponds and mud flats. They migrate throughout the East in spring and fall.

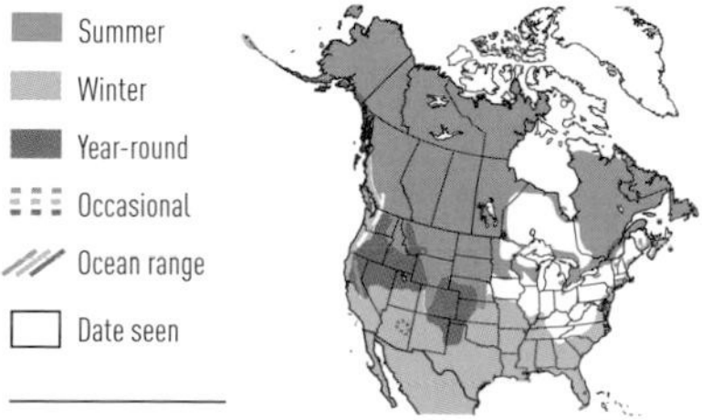

Location ________________

Anas discors Length: 15"

LOOK FOR: Named for the light blue wing patch that both sexes show in flight, the Blue-winged Teal is a delicate-looking small duck. The male's crescent-shaped white face patch on a blue-gray head is distinctive among our common ducks. Females are plain overall and are best identified by their shape rather than by any obvious field marks.

LISTEN FOR: Blue-winged Teal males utter a high, peeping whistle. Females give a froglike *querk*.

REMEMBER: The blue wing patches are not visible on swimming or resting birds, but if you scan a flock of ducks on a pond, you may be able to pick out the teal by their small size.

▼ *A wintering flock of Blue-winged Teal in a mangrove swamp.*

WOW!

Summer Teal is another name for this species because it seems to time its migration to avoid cold weather, migrating earlier in fall and later in spring than other ducks.

FIND IT: The Blue-winged Teal prefers freshwater ponds and marshes but can be found on almost any body of water, especially in migration. They migrate late in spring and early in fall. Most winter in Latin America.

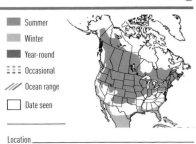

Summer
Winter
Year-round
Occasional
Ocean range
Date seen

Location ____________

LOOK FOR: Adult male is rich cinnamon red overall with a demonic-looking red eye. Female is similar to female Blue-winged Teal but plainer, especially in the face. This teal species is also notable for its larger bill, which is similar in shape and size to a shoveler's schnoz. In flight, Cinnamon Teal shows large white underwing patches; male shows bright blue shoulder patches.

LISTEN FOR: Male gives whistly peeps; female quacks nasally, sounding very similar to female Blue-winged Teal.

REMEMBER: A male Cinnamon Teal's bright coloration may be hard to miss, but females can go unnoticed. If you see a very plain female teal among a flock of puddle ducks, check to see if her bill is noticeably larger. If so, you can impress your friends by pointing out a female Cinnamon Teal!

WOW!

It is *not* true that the spice we call cinnamon comes from ground-up male Cinnamon Teal. That would not be cool at all.

◀ *Cinnamon Teal use their large bills to strain food from shallow water.*

FIND IT: Like other teal, the Cinnamon Teal prefers shallow water in ponds, marshes, sloughs, and flooded fields. Cinnamons are western birds and are rare east of the Great Plains.

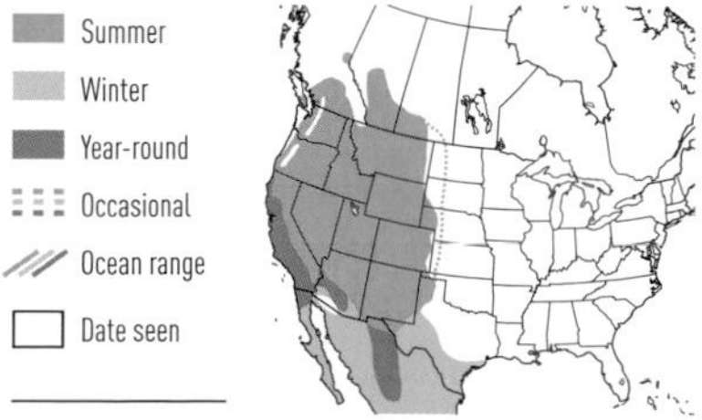

Location ______

GREEN-WINGED TEAL

Anas crecca Length: 14"

LOOK FOR: The Green-winged is our smallest teal and one of our smallest ducks. The male's handsome breeding plumage holds several key field marks: a dark rust-colored head with a swoop of green through the eye, a vertical white bar on the gray side, and a prominent horizontal patch of custard yellow below the tail, visible at great distances. Females show this custard patch too but are otherwise a rich brown overall.

LISTEN FOR: Males give a high *peep!* Females utter high, nasal quacks in a series.

REMEMBER: Flocks of Green-winged Teal are fast fliers, and small flocks maneuver in unison with amazing skill. Distant birds can often be identified by their small size.

WOW!

Other names for the Green-winged Teal include Common Teal, Mud Teal (it often forages on mud flats), Red-headed Teal, and Winter Teal (it winters farther north than other teal species).

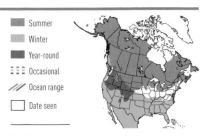

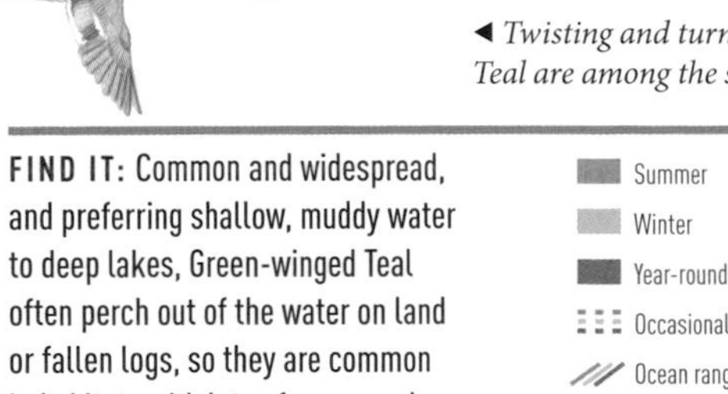

◀ *Twisting and turning in flight, Green-winged Teal are among the swiftest of ducks.*

FIND IT: Common and widespread, and preferring shallow, muddy water to deep lakes, Green-winged Teal often perch out of the water on land or fallen logs, so they are common in habitats with lots of snags and vegetation.

Summer

Winter

Year-round

Occasional

Ocean range

☐ Date seen

____

Location ______________

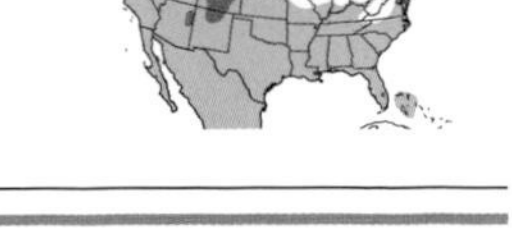

NORTHERN SHOVELER

Anas clypeata Length: 19"

LOOK FOR: From a distance the Northern Shoveler looks like a Mallard with a big nose. It's this bird's large, shovel-like bill that earned it its name. The male has a solid dark green head, large black bill, white chest, and rusty sides. The female is plain light brown overall with a large orange and gray bill. In flight, the male shoveler's huge light-blue wing shoulder patch is obvious.

LISTEN FOR: Males give a hollow-sounding *took-took,* uttered in pairs.

REMEMBER: At a distance, the best field mark for the shoveler is the huge rusty flank (side) patch surrounded by white.

WOW!

Imagine having to strain all of your food out of muddy water. The shoveler's bill is a huge filter with specially adapted comblike teeth that strain out tiny organisms—delicious!

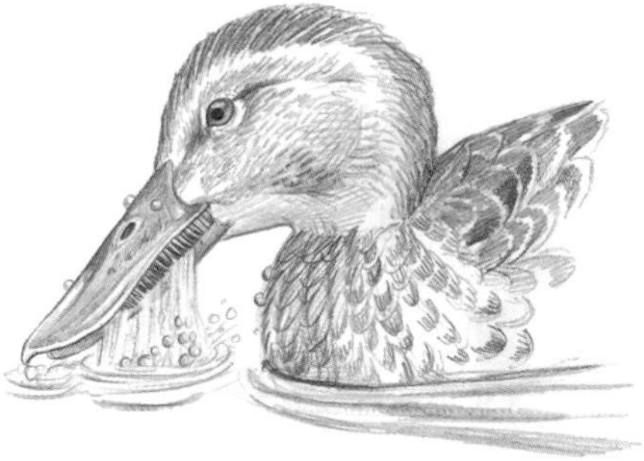

▶ *A female shoveler straining food items out of a billful of water.*

FIND IT: Shovelers prefer shallow water, such as ponds and marshes, where they swim with their huge bills submerged as they strain food from the muddy water. During migration, they can be found on almost any body of water.

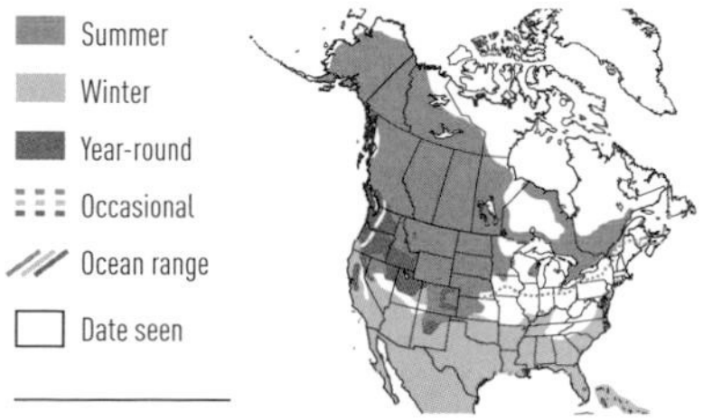

☐ Date seen ____________

Location ____________

Melanita perspicillata Length: 20"

LOOK FOR: Of our three scoter species, it is the Surf Scoter that is most easily identified. The male is black bodied with large white patches on the forehead and back of the head. In breeding plumage, the large orange and white bill of the male stands out. Adult females are dark brown with splotchy white patches on the head and face.

LISTEN FOR: Usually silent. Females give some croaks. Male's wings whistle in flight. Few birders hear these sounds.

REMEMBER: From a distance, Surf Scoters look bulky and dark, with big heads and large bills. The adult male's white head patches are distinctive.

WOW!

Duck hunters refer to the Surf Scoter as the Skunk-head Coot for the male's striking black-and-white head pattern.

▸ *Surf Scoters fly in wavering lines low over the water.*

FIND IT: Because they nest in the far North, Surf Scoters are most familiar to birders as wintering sea ducks along both coasts and on the Great Lakes. Large, wavering dark lines of scoters can be seen flying low over the water.

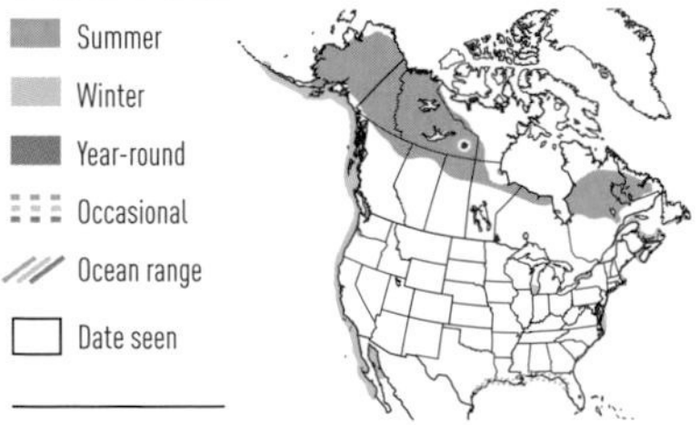

Location ____________________

REDHEAD

Aythya americana Length: 20"

LOOK FOR: A medium-sized diving duck named for its deep rust-red head. The male's black breast contrasts with a pale gray back. His gray bill is tipped in black, and the eyes are yellow. Female is uniformly gray-brown overall with a pale eye-ring and a pale area surrounding the base of the bill.

LISTEN FOR: Quacks like, well, a duck. Also makes some catlike mewing sounds.

REMEMBER: Very similar to the larger Canvasback, which has a bright white back. In mixed flocks of these two species, Redheads look dull by comparison.

▶ *Redheads are excellent divers. They get much of their food by diving for submerged aquatic vegetation.*

WOW!

Redheads are most active at night, spending days resting on the water. Imagine diving for your food at night!

FIND IT: More widespread in winter, when flocks can be found on open water, especially lakes and freshwater and saltwater bays. Breeds in freshwater marshes throughout the Great Plains to Alaska. Winters across the southern half of the U.S.

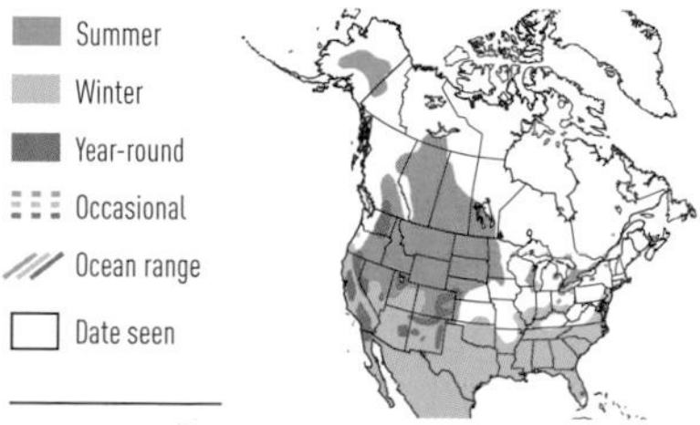

Location ______________________

Aythya collaris Length: 17"

LOOK FOR: This small peak-headed duck could be named the Ring-billed Duck for the white and gray rings on the male's black-tipped bill. Instead, it is named for the rarely seen subtle ring around the male's neck. A male in breeding plumage has a black head and back, and gray sides separated from the black chest by a bar of white. The female is brown-gray overall but has a white eye-ring and the same peaked head shape as the male.

WOW!

The Ring-necked Duck is named for a brown neck ring visible only on a bird in the hand. This is a holdover from the era of shotgun ornithology, when birds were shot, then examined, then named.

LISTEN FOR: Mostly silent. Females give a nasal, burry *errr-errr*. Males in courtship utter a high whistle.

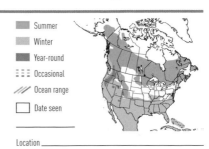

REMEMBER: Ring-necked Ducks appear very similar to both Greater and Lesser Scaup. But scaup have all-white sides while the Ring-neck's flanks are gray. And scaup have plain blue-gray bills.

◀ *With its peak-headed profile, the Ring-necked Duck may have been the inspiration for the cartoon character Daffy Duck.*

FIND IT: Look for Ring-necks on small wooded ponds at all seasons, often mixing with other diving duck species.

Summer
Winter
Year-round
Occasional
Ocean range
Date seen ________

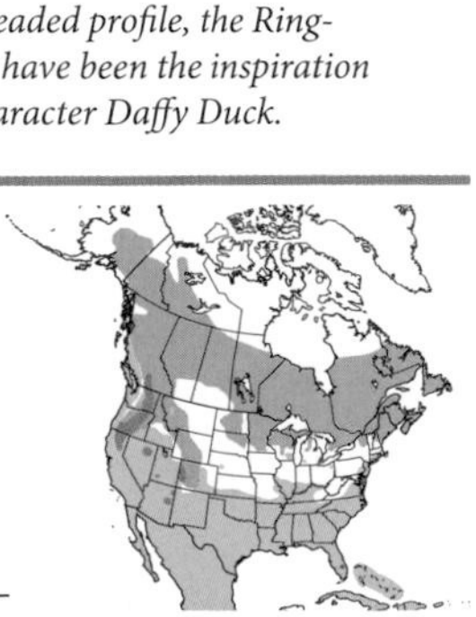

Location ________

LESSER SCAUP

Aythya affinis Length: 16½"

LOOK FOR: Of our two scaup species, the Lesser is the one more commonly seen inland. Male Lesser Scaup are dark headed and light backed. The female has a brown head with a white ring around the base of the bill. Lesser Scaup of both sexes can be separated from Greaters by head shape: the Lesser has a less rounded head, with a peak at the back. The Greater has a slight peak at the front.

WOW!

Both scaup species are known as Bluebills for the blue-gray bill color. The name scaup is derived from scallop, referring to the shellfish these ducks eat.

LISTEN FOR: Mostly silent. Female gives a loud, raspy *grr-grr*. Male whistles in courtship (not like *that!*).

REMEMBER: From a distance, the bright white sides and light back of scaup are distinctive. Telling scaup apart by head color is not reliable.

◄ *A female Lesser Scaup surfaces with a mussel. She will swallow it whole, and her stomach muscles will crush the shell.*

FIND IT: Lesser Scaup winter in huge flocks on inland lakes. Unlike other ducks, they are not often found in mixed-species flocks. When feeding, Lesser Scaup submerge with a diving plunge and bob back to the surface a few seconds later.

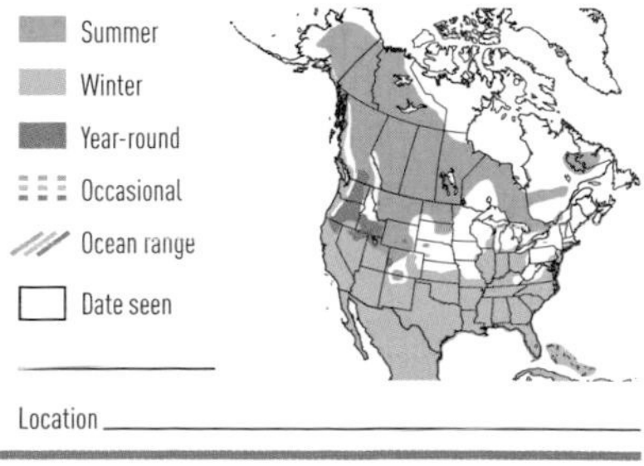

☐ Date seen

______

Location ______

Oxyura jamaicensis Length: 15"

WOW! Ruddy Ducks are also known as Stifftails for the male's spiky tail feathers, which are often raised up vertically while he is swimming.

LOOK FOR: The Ruddy Duck seems to have a big head and oversize bill for its body size. The rusty-bodied male in breeding plumage has a bright blue bill, dark head, and bright white cheeks, making him hard to mistake for anything else. The best field mark for the less distinctive female is her dirty white cheek divided by a horizontal line.

LISTEN FOR: Mostly silent except during courtship, when the male utters an otherworldly series of clicks perhaps best described as a water sprinkler that burps. The male Ruddy Duck also produces a variety of other sounds during courtship.

REMEMBER: Ruddy Ducks are not the best fliers. In fact, they will dive underwater to escape danger rather than take to the air.

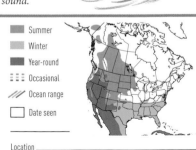

▶ *A courting male Ruddy Duck rapidly beats his bill against his breast, producing a peculiar drumming sound.*

FIND IT: A common duck found on ponds, marshes, bays, and lakes, the Ruddy Duck rarely mixes with other species in a flock. Huge winter flocks occur on lakes throughout the southeastern and south-central U.S.

Summer
Winter
Year-round
Occasional
Ocean range
Date seen ____

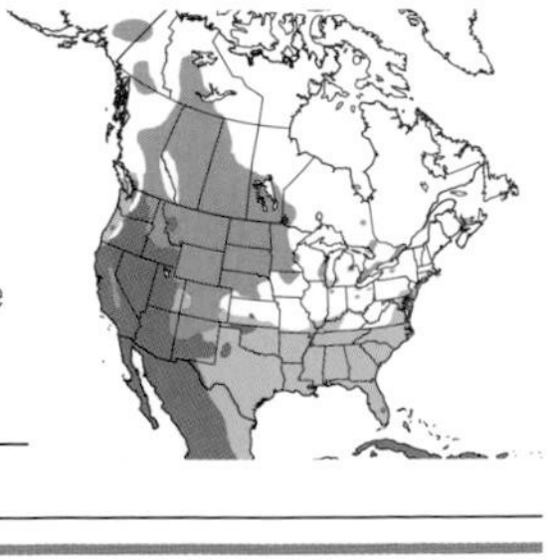

Location ____

LOOK FOR: Our largest merganser, the male Common Merganser appears mostly snow white from a distance, with a dark green head, long, slender orange bill, and black upper back. The female is gray bodied with a sharply contrasting rusty head, a white throat, and a shaggy crest.

WOW!

A Common Merganser scans for fish with its head underwater. When it sees one, it dives after it, propelling itself with its feet. The sharp, teethlike projections on its bill hold the slippery prey tight.

LISTEN FOR: The female gives a deep frog-like croak, *uhhnk-uhhnk*. The courting male sounds like someone gargling mouthwash. He also calls while swimming in splashy circles around the female.

REMEMBER: The long white body and long slender bill of the male Common Merganser are good field marks separating this species from other male green-headed ducks.

▶ *The serrated edges of the Common Merganser's bill are not teeth, but they function like teeth for holding slippery fish.*

FIND IT: The Common Merganser prefers large, open bodies of fresh water, especially lakes and large rivers. Wintering flocks on ice-free water can be large. Even at great distances, the male's low-slung white body is distinctive.

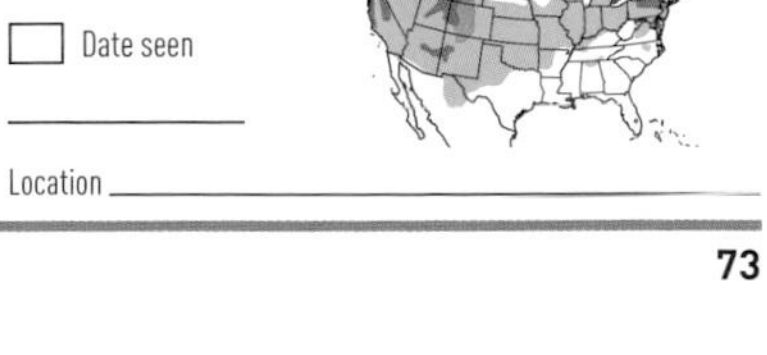

Location ______

RED-BREASTED MERGANSER

Mergus serrator Length: 23"

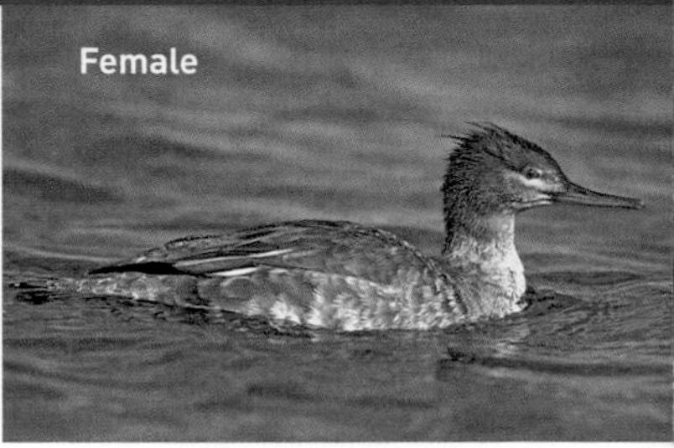

LOOK FOR: Our medium-sized merganser is the one with the shaggiest crest. Males have a white collar (a good field mark on distant birds) and the reddish breast for which they are named, a bright orange bill, and red eye. Females are similar to female Common Mergansers but are less clearly marked—there is no distinct dividing line between rusty head and gray body.

WOW!

Red-breasted Mergansers are fast, low fliers and are surprisingly fast swimmers too. With powerful strokes from their feet, they can pursue and catch the speediest of fish in shallow water.

LISTEN FOR: Usually silent. Female utters a harsh *kerr-kerr.*

REMEMBER: From a distance, the Red-breasted Merganser male looks much darker than the male Common Merganser. The Red-breasted's dark head and breast are separated by an obvious bright white collar.

◀ *Courting male Red-breasted Mergansers go into a frenzy when a female swims up—even in the dead of winter.*

FIND IT: Red-breasted Mergansers are most commonly found in migration on large bodies of water, but they prefer to winter in shallow saltwater bays along the coasts.

Summer

Winter

Year-round

Occasional

Ocean range

Date seen

Location

HOODED MERGANSER

Lophodytes cucullatus Length: 18"

LOOK FOR: Our smallest merganser species. Even from a distance, the male Hooded Merganser's bright white, fanned head crest outlined in black is easy to spot. Even when his crest is not fanned out, the bold black-and-white pattern remains visible. The female has the same big-headed, thin-billed appearance but with a body in tones of rusty brown.

WOW!
The Hooded Merganser can change its head shape depending on whether its crest is raised or lowered.

LISTEN FOR: Displaying males give a low, croaky *how loooong!* Females give a short, froglike croak. In flight, the Hooded Merganser's wings produce a high whistle.

REMEMBER: The male Hooded Merganser is similar to the male Bufflehead, but the Hooded has the crest outlined in black, orange flanks, and the two black vertical bars on the sides of the breast.

▸ *A female Hooded Merganser has a duckling ready to leave the nest in a natural cavity in a maple tree.*

FIND IT: Hoodies are usually seen in pairs or in small flocks. They prefer small wooded ponds, lakes, and swamps. This habitat, and their tendency to dive to avoid danger rather than take flight, makes them easy to overlook.

Summer
Winter
Year-round
Occasional
Ocean range
Date seen ________

Location ________

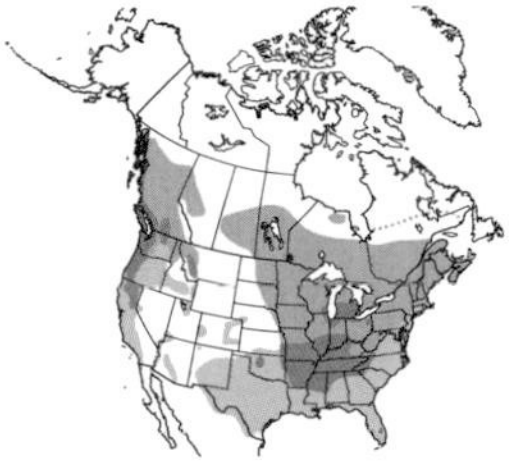

BUFFLEHEAD

Bucephala albeola Length: 13½"

WOW!

The Bufflehead is a cavity nester that is small enough to nest in holes excavated by Northern Flickers. How cool that a duck can nest in a hole made by a woodpecker!

LOOK FOR: This small peak-headed duck could be named the Ring-billed Duck for the white and gray rings on the male's black-tipped bill. Instead, it is named for the rarely seen subtle ring around the male's neck. A male in breeding plumage has a black head and back, and gray sides separated from the black chest by a bar of white. The female is brown-gray overall but has a white eye-ring and the same peaked head shape as the male.

LISTEN FOR: Mostly silent. Females give a nasal, burry *errr-errr*. Males in courtship utter a high whistle.

REMEMBER: Ring-necked Ducks appear very similar to both Greater and Lesser Scaup. But scaup have all-white sides while the Ring-neck's flanks are gray. And scaup have plain blue-gray bills.

▸ *The Bufflehead gets its name from the male's large-headed appearance, like a buffalo.*

FIND IT: Buffleheads nest near ponds in the north woods but winter across a vast portion of the continent in a variety of watery habitats, from large lakes to saltwater bays. They do not often mix with other duck species.

Summer

Winter

Year-round

Occasional

Ocean range

☐ Date seen

____

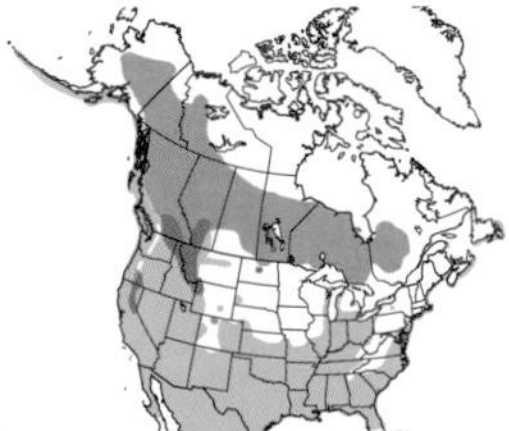

Location ______________________________

WILD TURKEY

Meleagris gallopavo Length: 37" (female) to 46" (male)

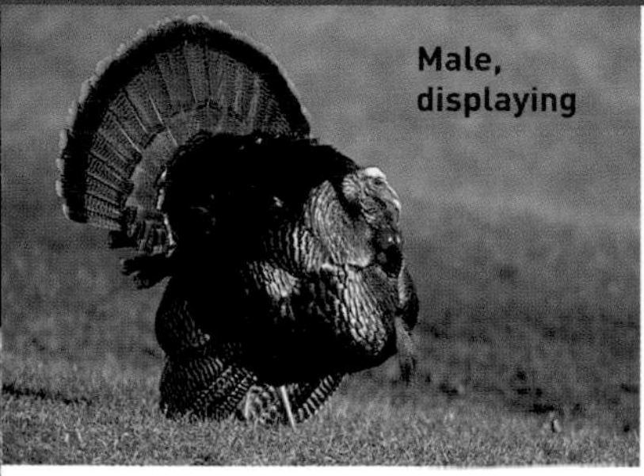

LOOK FOR: The Wild Turkey is a surprisingly large bird, with a featherless head and all-dark body perched atop stout pink legs. The dark body feathers can show shades of green, orange, and blue in direct sunlight, and the bare head skin also changes color, especially on males during their tail-fanning courtship display. Females are smaller, with gray, not pink, heads.

WOW!
The Wild Turkey can fly surprisingly fast for such a large bird, but only for short distances. In flights of less than a mile, it may reach speeds of 55 to 60 miles per hour.

LISTEN FOR: Displaying males give a loud, distinctive gobble. Females call their young (known as poults) with a sharp *tuk! tuk!*

REMEMBER: The Wild Turkey's huge size, bald head, and dark coloration are distinctive.

▶ *Young male Wild Turkeys have shorter "beards" hanging from their chests.*

FIND IT: The Wild Turkey seems to prefer mixed woods that contain acorn-producing oaks. Spring flocks may hold courtship displays in open fields or woodland meadows. Flocks roost in trees at dawn and dusk.

Summer
Winter
Year-round
Occasional
Ocean range
Date seen

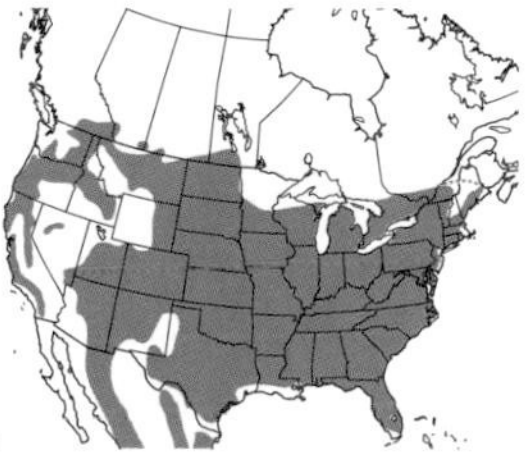

____

Location ______________

Bonasa umbellus Length: 17"

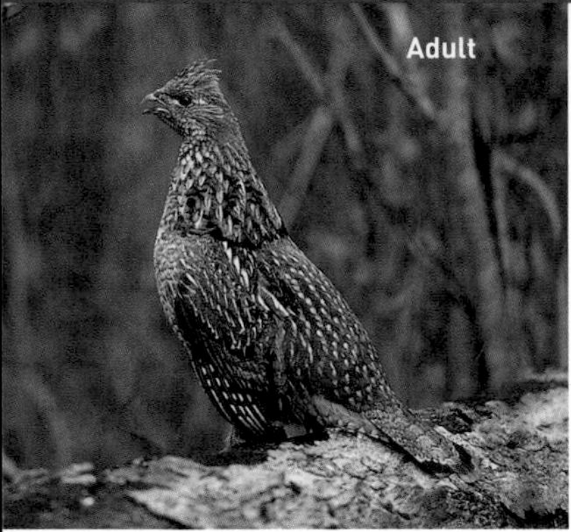

LOOK FOR: This game bird is decked out in browns, grays, and blacks that are perfect camouflage for its woodland habitat. Sexes are similar. The Ruffed Grouse occurs in two plumage varieties: gray and red (rusty orange).

LISTEN FOR: The male Ruffed Grouse performs a drumming display, beating his wings against the air while perched on a log, deep in the woods. This sounds like a muffled engine starting up: *whup-whup-whup-whup*, slowly at first, then faster and faster until it dies off.

REMEMBER: Ruffed Grouse tend to stay still for as long as possible as danger approaches, relying on their cryptic coloration to conceal them.

WOW!

In spring, territorial male Ruffed Grouse are surprisingly aggressive. They have been known to challenge (and peck at) hikers, bicyclists, and even cars driving on roads through their territories.

▶ *Ruffed Grouse are named for the male's black neck ruff, which he puffs out during his drumming display.*

FIND IT: Widespread in woodland settings, this is our most easily encountered grouse species. During spring and summer, locate males by their drumming sounds. In summer, you may see females leading chicks across woodland clearings.

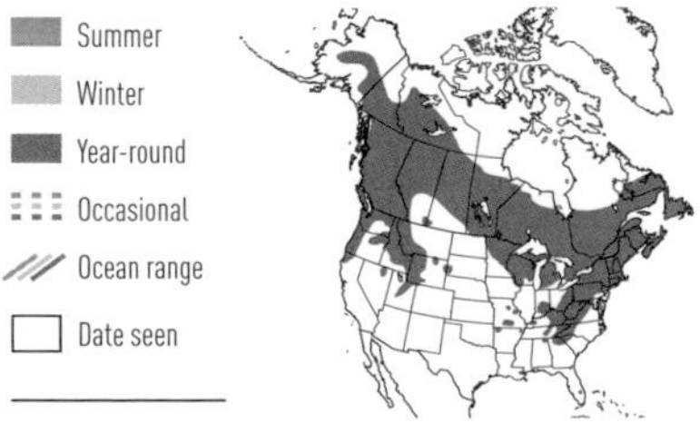

Date seen

____

Location ______________________________

Dendragapus obscurus, Dendragapus fuliginosus Length: 20"

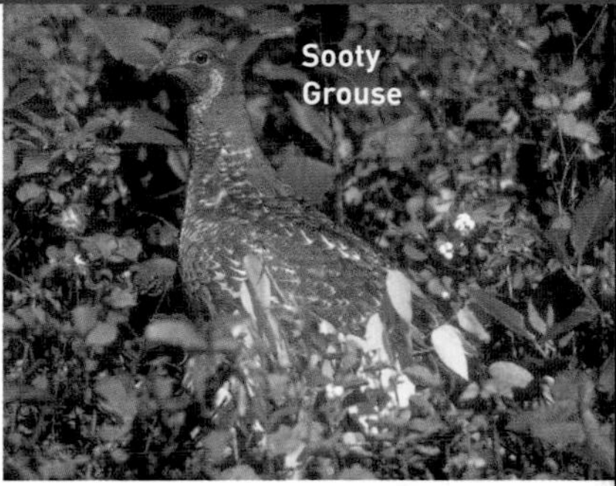

LOOK FOR: Until 2006 these two large dark grouse species were a single species, the Blue Grouse. Males of both species are dark bluish brown with bright yellow eye combs and neck sacs, which can be inflated during courtship displays. Females are dark gray, perfect cryptic coloration for their woodland habitat.

WOW!
Both species will freeze in place when disturbed, in the hopes that their cryptic coloration will conceal them.

LISTEN FOR: Males of both species give a series of low-pitched whoops during courtship displays.

REMEMBER: Don't be fooled by the photos of displaying male grouse. It's fairly rare to see the males with tails fanned and neck sacs inflated. Use range to help tell these two species apart.

▶ *These birds survive in winter on a diet consisting almost exclusively of conifer needles. Yum!*

FIND IT: Both species are year-round residents of mature woodlands. Dusky prefers more open habitat, including burned areas and meadows. Its range covers more interior montane forests. Sooty inhabits semi-open fir and spruce woods, and its range is closer to the Pacific Ocean.

☐ Date seen ______________ ☐ Date seen ______________

Location ______________ ______________

RING-NECKED PHEASANT

Phasianus colchicus Length: 21" (female) to 32" (male)

LOOK FOR: A large long-tailed bird most often seen walking on the ground, the Ring-necked Pheasant is not as shy as other game birds. The male is distinctive with his rusty body, white neck ring, red face, and showy tail feathers. Females are the same shape but are tan overall and have shorter tails.

LISTEN FOR: Only the males call, uttering a harsh, two-syllable *craa-cahh!* Pheasants' wings make a loud ruffling sound as the birds burst into flight from a concealed spot.

▲ *The rough double crow of the Ring-necked Pheasant is a signature sound of the prairie spring.*

REMEMBER: It's hard to confuse a male Ring-neck with any other bird. To help separate the drab female from prairie grouse species, look for her long tail and elongated body shape.

WOW!

The Ring-necked Pheasant was introduced to North America from Asia as a game bird. In many areas, the Ring-necked Pheasant population is supplemented by the birds raised on game farms.

FIND IT: Widespread in open farmland habitat, Ring-necks prefer old-fields, grain fields, marshy grasslands, and hedgerows. They can often be seen foraging by field edges along rural roads. They prefer to run from danger rather than fly.

Summer
Winter
Year-round
Occasional
Ocean range
Date seen

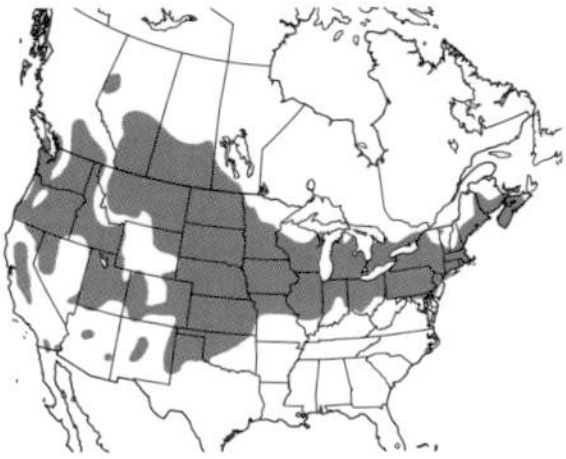

Location ______

CALIFORNIA QUAIL and GAMBEL'S QUAIL

Callipepla californica, Callipepla gambelii Length: 10", 10½"

LOOK FOR: These two smallish chickenlike birds appear very similar from the wings up. Males and females have distinctive teardrop-shaped black head plumes. Males of both species have black faces outlined in white. The unscaled butter yellow belly is diagnostic of the male Gambel's, while the male California's belly is scaled with dark edges on tan feathers. Females are dull gray-brown overall.

WOW!
One day after hatching, downy quail hatchlings leave the nest and follow their parents. These tiny tots are supercute!

LISTEN FOR: California's three-syllable call is *Chi-CAH-go!* Also gives *clucks*, *pits*, and a loud *krrrr!* Gambel's main vocalization is a loud, descending *kaaah!* Also a four-part *kah-KAH-kah-kah*.

▲ *While foraging, coveys of both of these quail species will post a lookout to watch for approaching danger.*

REMEMBER: It can be hard to remember which quail species is which, despite their limited range overlap. Two ways to distinguish them: the Gambel's Quail is sometimes called "redhead" by hunters, and they "gamble" with their lives by living in drier desert habitat.

FIND IT: Both are nonmigratory residents. California Quail are more coastal in distribution and prefer greener, more vegetated habitat, including chaparral, coastal scrub, woodland edges, and parks. Gambel's prefer drier desert habitat, including canyons, and open areas with scattered brush.

☐ Date seen ______________ ☐ Date seen ______________

Location ______________ ______________

Colinus virginianus Length: 10"

Female (left), male (right)

LOOK FOR: The masked appearance of the male Northern Bobwhite is the best visual field mark for this species, but this quail is often heard before it is seen. The female has a pale tan version of the male's mask. Populations of Northern Bobwhites are declining in many areas.

LISTEN FOR: The Northern Bobwhite says its own name—at least the *bobwhite!* part. In spring and summer, males give this call repeatedly from an exposed perch. Members of a quail flock, or covey, utter a variety of whistling calls.

REMEMBER: To avoid danger, bobwhites usually run for cover and will fly only when surprised or when running is not a safe option.

WOW!

To guard against predators at night, Northern Bobwhite coveys settle down in a circle with all birds facing out. If one bird sees a predator, it explodes into flight, which tells the nearby birds to do the same.

▶ *Bedding down in a tight circle, each bird facing out, a covey of bobwhites prepares for the night.*

FIND IT: Listen for the distinctive *bobwhite!* call in old pastures, farm fields with hedgerows, open-understory woods, and grasslands, and scan nearby perches such as low snags and fenceposts for the one bird that acts as the sentinel for the covey.

Summer

Winter

Year-round

Occasional

Ocean range

Date seen ____________

Location ____________

COMMON LOON

Gavia immer Length: 32"

LOOK FOR: The Common Loon is a large, long diving bird seen most often on big lakes, where it rides low in the water. In summer, the adult's black head, body, and bill contrast with a bright white chest and black-and-white-checkered back. In winter, Common Loons are drab gray above and white below.

WOW!

Hollywood movie directors often dub the call of the Common Loon into their movies. Too often, however, they insert it in the wrong setting, such as in the desert or in a junkyard in New York City!

LISTEN FOR: The haunting high yodel of the Common Loon echoes across wooded lakes in the North. It also gives a laughing *ha-ha-ha-ha* call.

REMEMBER: You may confuse a distant Common Loon on the water with a Double-crested Cormorant. The loon has a heavier overall appearance; a thicker, shorter neck; and a heavier, dark bill.

▲ *Rearing up like a cobra, a male Common Loon defends his nest from an intruder.*

FIND IT: Summer finds Common Loons on large, quiet bodies of water. They winter on waters along both coasts and fly in small flocks low over the water. During migration, loons may be found on large rivers, lakes, and reservoirs.

Summer
Winter
Year-round
Occasional
Ocean range
Date seen ______

Location ______

Podilymbus podiceps Length: 13"

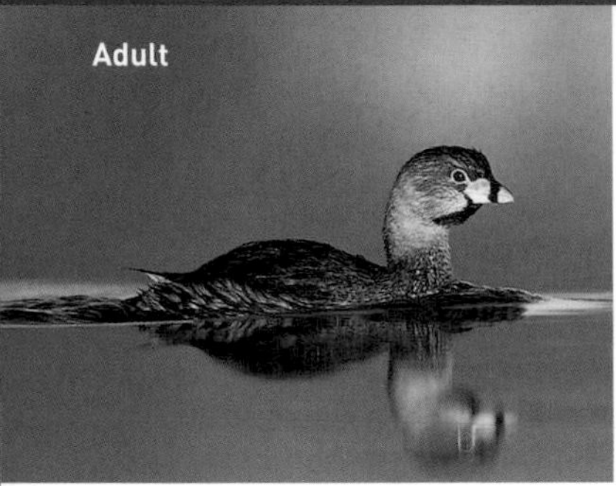

LOOK FOR: Our most widespread and common small grebe, the Pied-billed is usually seen alone (not in flocks) swimming on quiet waters where it dives and pursues small fish underwater, propelled by its feet. Tawny brown overall in all seasons, it gets its name from the black band around its pale bill, a field mark present only in summer.

LISTEN FOR: For such a small bird, the Pied-billed Grebe has a big voice, giving a long, rapid-fire series of grunts, toots, and hoarse barks.

REMEMBER: The Pied-billed's brownish overall color, thick bill, and all-dark eyes set it apart from our other small grebes.

WOW!

To escape danger, the Pied-billed will submerge and swim a long distance underwater. It may then raise only its head above the water, watching quietly for the danger to pass.

▶ *A Pied-billed Grebe can compress its plumage, exhale, and sink without a ripple.*

FIND IT: Almost any North American pond or lake can host Pied-billed Grebes, but they prefer water with concealing vegetation. The Pied-billed Grebe's small size, drab coloration, and solitary nature make it easy to overlook.

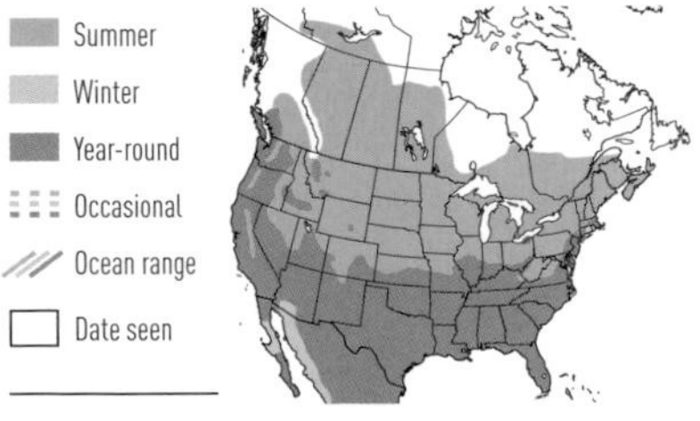

Location ______________________

HORNED GREBE

Podiceps auritus Length: 14"

LOOK FOR: The Horned Grebe is named for the golden yellow patch of feathers on the sides of its dark head. You won't see this on birds in winter, when they are dark gray and white. The Horned Grebe's white face and neck help to set it apart in winter. Breeding-plumage bird has a black crown, throat, and back with a rusty neck and sides.

WOW!

In all seasons the Horned Grebe has bright red eyes. This gives birds in colorful breeding plumage an especially freaky look.

LISTEN FOR: Mated pairs give a high-pitched twittering duet. In summer their other common call is a sputtering *pitpitPRE-ahhh!*

REMEMBER: Horned and Eared Grebes both have small, thin bills, but the Horned Grebe almost always has a white tip on its dark bill. The Pied-billed Grebe has dark eyes and a stocky bill.

▲ *A lucky birder might see a Horned Grebe in breeding plumage, but winter plumage (right) is more commonly seen.*

FIND IT: Look for Horned Grebes swimming, diving, and nesting on freshwater lakes. Many spend the winter in coastal (saltwater) bays. Others winter on reservoirs and big lakes. During migration they can be found throughout the East.

Summer
Winter
Year-round
Occasional
Ocean range
Date seen

Location

WESTERN GREBE and CLARK'S GREBE

Aechmophorus occidentalis, Aechmophorus clarkii Length: 25"

LOOK FOR: By far our largest grebes, these two species are very similar: both are dark bodied with long white necks, black caps, and yellow bills. The black cap of the Western Grebe encloses the red eyes. On the Clark's Grebe the eye is surrounded by white. Clark's bill is brighter yellow.

LISTEN FOR: Western: loud, double *crik-crik*. Clark's: a single-note *creek*.

REMEMBER: The Western Grebe has dark surrounding its eyes, like sunglasses, guarding it from the "western" sun.

WOW!

One of these grebe species is among a handful of species discovered and named during the Lewis and Clark expedition in the early 1800s. Guess which one . . . *duh!*

◀ *Males of both of these two large grebe species perform an exciting courtship dance, running across the surface of a lake with their necks extended. This never fails to impress the ladies.*

FIND IT: Ranges and habitat preferences of the two species overlap: in all seasons, both can be found on large lakes and freshwater bays. Western Grebes commonly winter on salt water, while Clark's prefer fresh water all year.

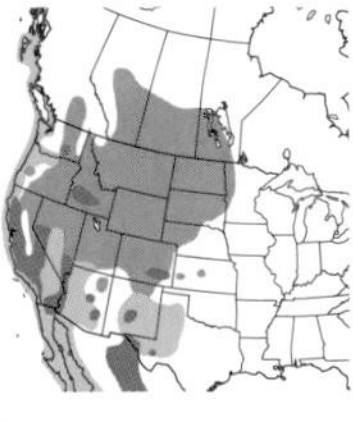

WESTERN GREBE

☐ Date seen ______________

Location ______________

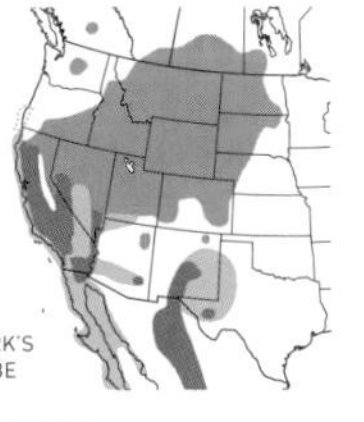

CLARK'S GREBE

☐ Date seen ______________

____

AMERICAN WHITE PELICAN

Pelecanus erythrorhynchos Length: 62"

LOOK FOR: This huge white bird (with a nine-foot wingspan!) has black flight feathers and a large yellow-orange bill, making it hard to mistake for any other bird. Often found in flocks foraging in shallow lakes as well as soaring high overhead.

LISTEN FOR: Not very vocal, except in breeding colonies, where they utter low, croaking grunts.

REMEMBER: The White Pelican differs in a few important ways from its cousin the Brown Pelican. The White Pelican forages in cooperative flocks, which herd schools of fish to waiting, open bills. The Brown Pelican dives for its food. White Pelicans are inland nesters, while Brown Pelicans are closely associated with coastal saltwater areas.

WOW!

A White Pelican's pouch can hold more than two and a half gallons of water. Imagine holding that much water in your throat. Now imagine all that water full of squirming fish. Yum, sushi!

▲ *A White Pelican guards its chick in the nesting colony.*

FIND IT: During spring and summer, White Pelicans breed in big colonies, usually on islands in large inland lakes. They range far from the colonies to forage on shallow lakes and marshes. They winter along the coasts.

Location ______________

BROWN PELICAN

Pelecanus occidentalis Length: 51"

LOOK FOR: Though it's not as large as the American White Pelican, the Brown Pelican is still a huge bird. Its mostly brown body, dusty gray wings, and yellow crown patch are minor field marks compared to the oversize pelican bill with its expandable pouch. Breeding-season adults have darker brown necks and more yellow on the heads.

LISTEN FOR: Except for young birds in the nesting colonies, these birds are nonvocal.

REMEMBER: Brown Pelicans are large-headed, large-billed, slow fliers. Soaring low over the water on their vast wings, they are hard to confuse with any other bird.

WOW!

Harmful chemicals nearly wiped out the Brown Pelican. In the early 1970s, only a small population remained. Thanks to the banning of DDT and other chemicals, Brown Pelicans have made a strong recovery.

▶ *A Brown Pelican fishes by folding its wings and plummeting, scooping up the fish in its enormous gular pouch.*

FIND IT: Common along both coasts, but fairly unusual inland or far from salt water, the Brown Pelican has expanded its range northward. Watch for them perched on pier pilings or in small groups soaring low over the water.

Summer
Winter
Year-round
Occasional
Ocean range
Date seen ______

Location ______

DOUBLE-CRESTED CORMORANT

Phalacrocorax auritus Length: 33"

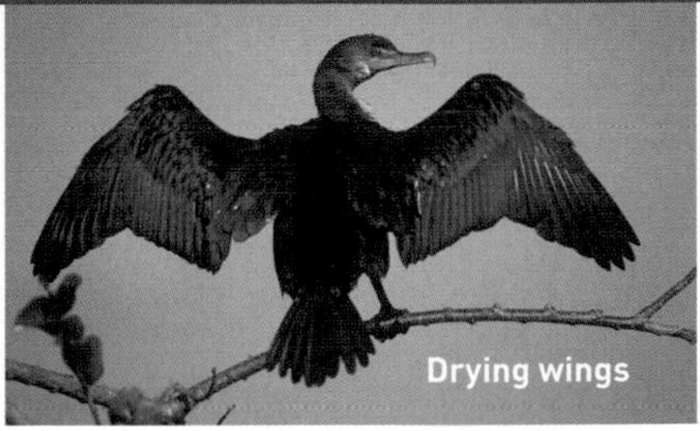

LOOK FOR: Our most common cormorant continent-wide, the Double-crested is a long and slender all-dark bird with a pale yellow-orange bill and throat patch. Its crests are apparent only in the breeding season, and even then they are difficult to see. Cormorants float low in the water and dive after fish, pursuing them underwater. Flocks of migrant cormorants fly overhead in wavering Vs. Young birds have pale breasts.

LISTEN FOR: Nonvocal away from nesting colonies.

REMEMBER: The Double-crested is the only cormorant commonly found on inland bodies of water. The Anhinga is similar but has many white feathers on its back and swims with only its head above water.

WOW!

In some areas, cormorant populations are controlled because of their perceived impact on populations of game fish.

▶ Although cormorants are often blamed for eating sport fish, the majority of their diet is fish unwanted by people.

FIND IT: Large gatherings of cormorants occur on large lakes, along rocky coasts, and on snags along rivers and bays. After diving for fish, Double-crested Cormorants frequently perch near the water and extend their wings to dry them out.

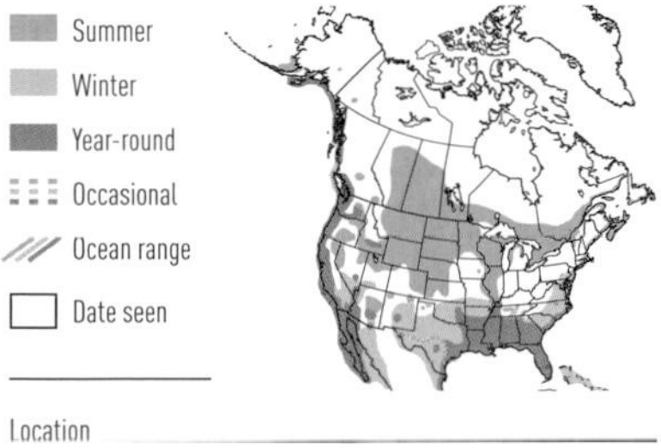

Anhinga anhinga Length: 35"

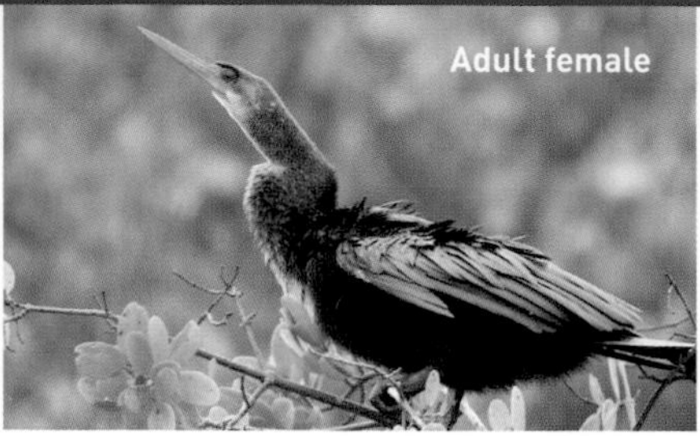

LOOK FOR: A long thin neck and small head, thin yellow bill, and white feathers on the wings and back set the Anhinga apart from the closely related cormorants. Anhingas commonly perch with their wings spread and swim with only their heads above water. Adult males are mostly black overall, while adult females and young birds have tan heads and necks atop black bodies.

LISTEN FOR: A series of rapid clicks that sound mechanical, like a sewing machine.

REMEMBER: Anhingas are excellent fliers, often soaring high in the sky. When it soars, the long, pointed wings and long tail and neck make the Anhinga look like a flying X.

▲ *An Anhinga spears a fish, tosses it into the air, and swallows it whole, headfirst.*

WOW!

The Anhinga is known by many names, including Snakebird (for its snakelike head), Water Turkey (for its fanned tail), and Black Darter (for its ability to spear fish with its sharp bill).

FIND IT: On slow-moving rivers, lakes, ponds, and swamps in the Southeast, Anhingas often perch on snags near or in the water, hanging their wings out to dry. Anhingas nest in colonies with other wading birds.

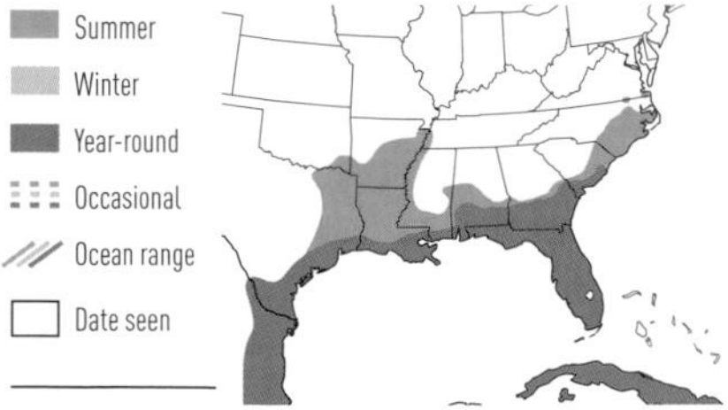

Date seen ______________

Location ______________

GREAT BLUE HERON

Ardea herodias Length: 46"

LOOK FOR: A tall, long-legged, grayish bird with a long, yellow, daggerlike bill used to spear fish and other prey. Great Blues are slow and stately in their movement, whether wading or flying. Flying birds show black wingtips, and the legs trailing behind give the bird a superlong appearance.

LISTEN FOR: A deep, croaking *graaaak*, usually given in flight, that sounds more like a belch than a bird's call.

REMEMBER: Other dark wading birds resemble the Great Blue but are much smaller, including the Green Heron, Tricolored Heron, Little Blue Heron, and the Glossy Ibis.

WOW!

Many Great Blues migrate south in winter, but a few may linger. When water freezes they may hunt in nearby fields for rodents and small birds.

► *Great Blue Herons nest in large colonies called rookeries. Their large, bulky stick nests are easy to spot in early spring.*

FIND IT: Widespread and increasingly common. Always found near water, especially clear, calm water where the Great Blues can easily see and hunt for fish, crayfish, frogs, and other prey.

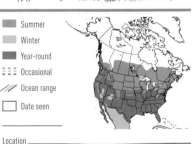

Summer
Winter
Year-round
Occasional
Ocean range
Date seen

______

Location ______

Adult. breeding

LOOK FOR: The Great Egret is a tall, slim, all-white bird with a long yellow bill and long all-black legs. It can appear extremely tall when the long neck is extended, or smaller when the neck is pulled in close to the body. Lacks the long showy plumes of other herons and egrets.

LISTEN FOR: Low, unmusical croaks and harsh squawks.

REMEMBER: Other white egrets are smaller. The Snowy Egret has black legs with yellow feet. The white-morph Reddish Egret has gray legs and a pink and black bill. The Cattle Egret is half the Great Egret's size and has a stubby bill by comparison.

WOW!

Great Egret plumes adorned women's hats in the late 1800s, and this nearly resulted in the species' extinction. Conservation efforts saved the egrets.

◀ *If you see a group of wading birds together, the long-legged Great Egret is likely to be in the deepest water.*

FIND IT: Our most common large white wading bird, the Great Egret is often seen standing completely still, watching for a passing fish, frog, or snake, on marshes, sloughs, and lakes. They may wander far north of their normal range in late summer.

Summer

Winter

Year-round

Occasional

Ocean range

Date seen ____________

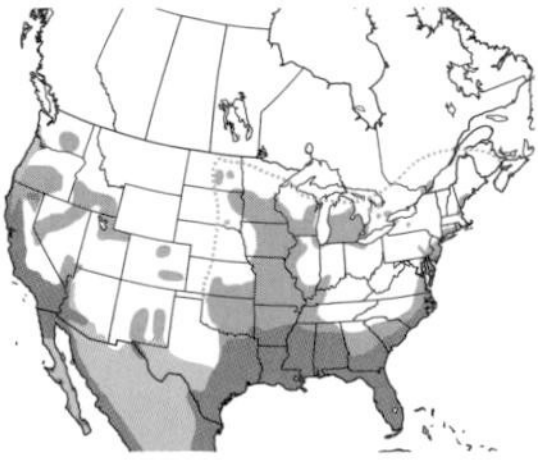

Location ____________

Egretta thula Length: 24"

LOOK FOR: Perhaps the best field marks for this all-white bird are its yellow "slippers"—golden feet at the bottom of black legs. Yellow also appears on the lore, the area between the eyes and the slender black bill. During the spring and summer, adult Snowies grow long, lacy plumes on their heads, necks, and tails. This is their showy breeding plumage.

LISTEN FOR: A series of gurgling croaks that sounds like someone is getting sick.

REMEMBER: The Snowy is the most delicate looking of our white egrets, with its fine-pointed bill, feathery plumes (in spring and summer), and overall slender appearance. It often forages with other species of wading birds.

WOW!

The Snowy Egret stirs up water with its feet, hoping to scare a minnow or crayfish out into the open so the egret can snatch the prey with a strike of its lightning-fast neck and bill.

◀ *By rapidly patting the water's surface with its foot, a Snowy Egret attracts fish and stirs up other food from the muck.*

FIND IT: Snowies prefer wide-open marshes, ponds, and shores—either freshwater or saltwater habitats will do. They wander north in late summer. In winter, northernmost birds move south to coastal areas.

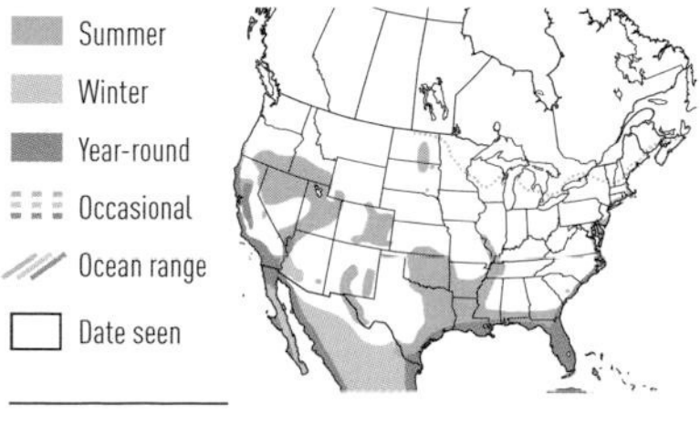

____

Location ______________________________

Bubulcus ibis Length: 20"

LOOK FOR: The Cattle Egret is the shortest, stockiest of our white egrets and herons, and its most reliable field mark is its stout yellow bill. In winter, the Cattle Egret's plumage is all white, but during the breeding season, an adult shows rusty patches on the head, breast, and back.

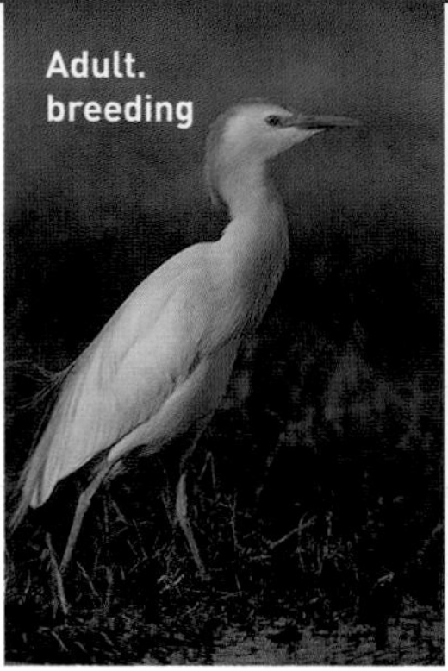

LISTEN FOR: Cattle Egrets are not very vocal birds, except in the nesting colony, where they utter a series of guttural quacks that sound more like a pig than a bird.

REMEMBER: A chunky, white wading bird in the middle of an open field is likely to be a Cattle Egret. A tall, slender, white wading bird standing in water is more likely to be a Great or Snowy Egret.

▼ *Cattle Egrets follow large grazing animals and even tractors, eating the insects frightened from the grass.*

WOW!

The Cattle Egret is native to Africa but has expanded to areas around the world. The first Cattle Egrets arrived in North America in the early 1950s and have since spread across much of the continent.

FIND IT: The well-named Cattle Egret can be seen foraging in fields and pastures near cattle and other livestock, eating insects disturbed by the animals' movement. Cattle Egrets nest in colonies, often with other herons and egrets.

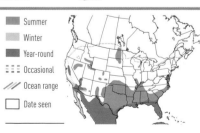

Summer
Winter
Year-round
Occasional
Ocean range
Date seen

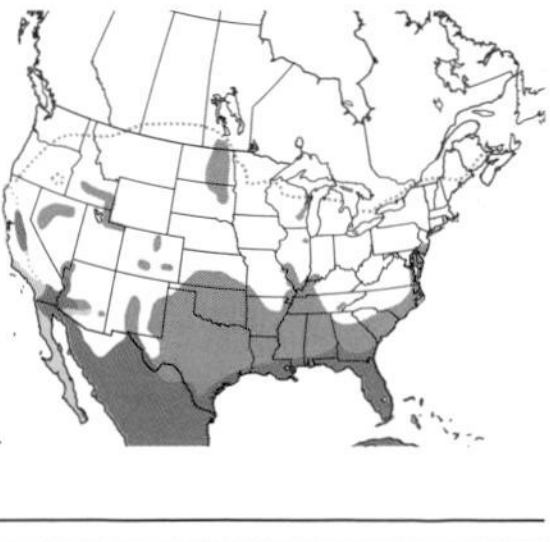

____

Location ______________

BLACK-CROWNED NIGHT-HERON

Nycticorax nycticorax Length: 25"

LOOK FOR: A chunky bird with a gray body, black crown and back, bright orange-red eyes, and sturdy yellow or pink legs. Black-crowned Night-Herons often perch or stand in a hunched-over pose. Crown and back color are black, differing from the closely related Yellow-crowned Night-Heron (adults of which have yellow crowns and gray wings and backs).

LISTEN FOR: A loud, low-toned *qwock!* often uttered in flight can identify Black-crowns flying unseen at night.

REMEMBER: Young night-herons of both species are streaky overall and look dif ferent from adults. The yellow in the bills of young Black-crowns helps to tell them from young Yellow-crowns.

▲ *The Black-crowned Night-Heron's enormous eyes allow it to find its prey in almost total darkness.*

WOW!

Black-crowned Night-Herons nest on every continent on earth, making them one of the few truly worldwide bird species.

FIND IT: During the day both Black-crowned and Yellow-crowned Night-Herons can be found roosting in trees in their nesting colonies. At night they leave the colonies to forage in nearby ponds, rivers, and marshes.

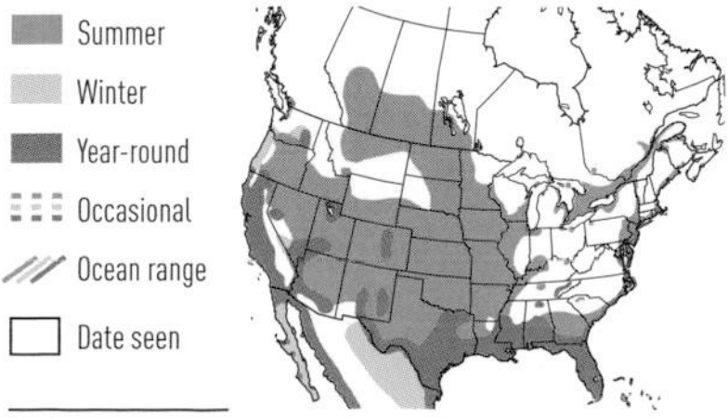

Location ____________________

LOOK FOR: This very small, dark heron has the perfect coloration to blend in with habitat along the water's edge: dark tones of reddish brown and deep green, with bright yellow-orange legs. Though its body is stocky, its bill is slender and long, perfect for spearing small fish, its preferred food.

LISTEN FOR: A loud, hoarse *skowp!* Often given when the bird is spooked into taking flight.

WOW!

Green Herons may use bait (such as bread or corn that's been fed to ducks, or even a piece of vegetation) to attract curious fish to within striking distance.

REMEMBER: Most easily confused with the night-herons and bitterns, but the Green Heron is much darker overall and shows all-dark wings in flight.

▶ *A Green Heron bait-fishing with a piece of bread.*

FIND IT: Quiet wooded streams, small ponds, and almost any body of fresh water that is lined with thick vegetation may host a Green Heron. Scan the trees and snags along the water's edge to find a Green Heron sitting quietly.

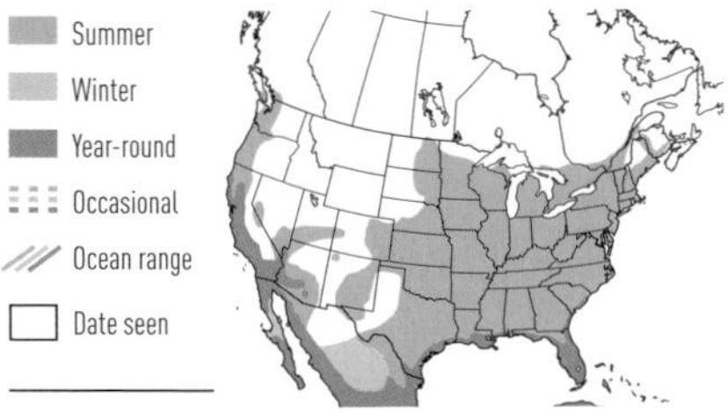

Location ______

LOOK FOR: A large, long-legged, dark-overall bird with a downward-curving bill, the Glossy Ibis gets its name from the glossy appearance of its plumage, with its shades of metallic green, black, purple, pink, and rust. Breeding birds show more rusty coloration on the neck and body.

LISTEN FOR: Not a very vocal species, but gives a nasal *uhn-uhn-uhn-uhn* call when flushed.

REMEMBER: The Glossy Ibis is hard to tell apart from the White-faced Ibis, which is common in summer in the Great Plains and western U.S. In breeding season, the Glossy has a white outline on a dark face while the White-faced has a white outline on a red face.

WOW!

The Glossy Ibis finds its food by feeling it with the tip of its long bill. As it probes in the mud of a shallow marsh, it senses a food item (such as a beetle or other insect) and snaps it up.

▶ *Glossy Ibises probe deep in the mud. When something moves, they grab it with the tip of their sickle-shaped bill.*

FIND IT: The Glossy Ibis is commonly found in marshes and wetlands. They are almost always found in flocks, wading in shallow water and probing in the mud for food. This gives them a hunched-over look, like workers picking vegetables in a field.

Summer

Winter

Year-round

Occasional

Ocean range

Date seen

____

Location ______________

Eudocimus albus Length: 25"

LOOK FOR: The adult White Ibis is an impressive bird: all-white body, black wingtips, red legs, and down-curved red bill set off with a blue eye. Young birds are a mixture of brown and white but still show the red bills and legs. In flight, the black wingtips stand out on white wings.

WOW!

Nesting in colonies in trees, both males and females share the nest-building chores, with the male bringing the materials and the female doing the actual building.

LISTEN FOR: A dull-sounding *uhn!* Most commonly heard from the nesting colonies.

REMEMBER: Even viewed at a great distance, the decurved shape of the bill is an excellent field mark to separate this species from other large white wading birds.

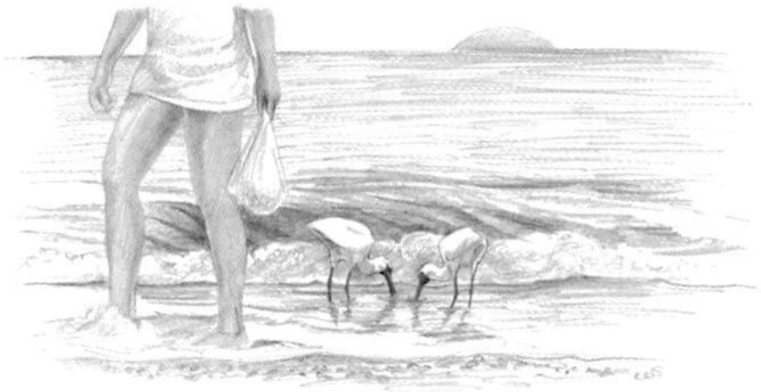

▶ *Unlike other ibises, the White Ibis is comfortable sharing its habitat with humans, especially in Florida.*

FIND IT: Common in a variety of watery habitats, White Ibises prefer to be in groups, and small flocks are commonly seen foraging for insects on beaches, parks, golf courses, and large expanses of suburban lawn.

Summer
Winter
Year-round
Occasional
Ocean range
Date seen

____

Location ______________

ROSEATE SPOONBILL

Platalea ajaja Length: 32"

WOW!
One of the most beautiful sights a birder can see is a flock of Roseate Spoonbills at sunset flying back to the nesting colony.

LOOK FOR: The Roseate Spoonbill is hard to confuse with anything else. Its rose pink coloration is stunning, and its spoon-shaped bill is unlike that of any other North American bird. Young birds are light pink, attaining the more intense adult colors by their third year. An up-close look at its massive bill, bald head, and red eye might convince you that this bird looks best from a distance.

LISTEN FOR: Not very vocal but does utter a series of low grunts on a single pitch.

REMEMBER: If you see a flock of large pink birds along the coast, from Florida to Texas, they are almost certain to be spoonbills.

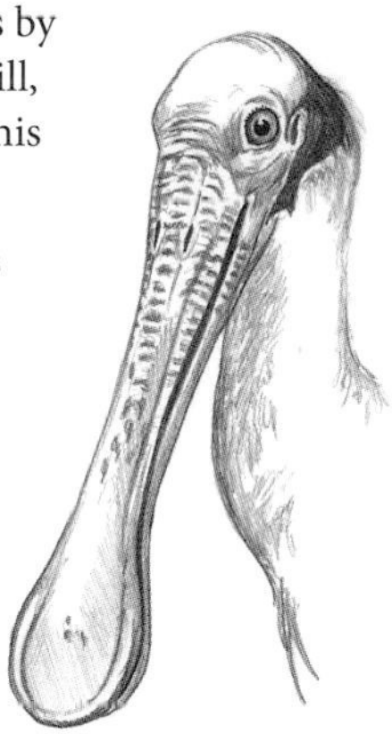

► *The spoonbill's bill is full of nerve endings, allowing it to feed by "feel."*

FIND IT: Common within its coastal range in shallow bodies of both fresh and salt water, where it forages by sweeping its head back and forth and snapping its bill shut when it feels a prey item—usually a small fish.

Date seen ____________

Location ____________

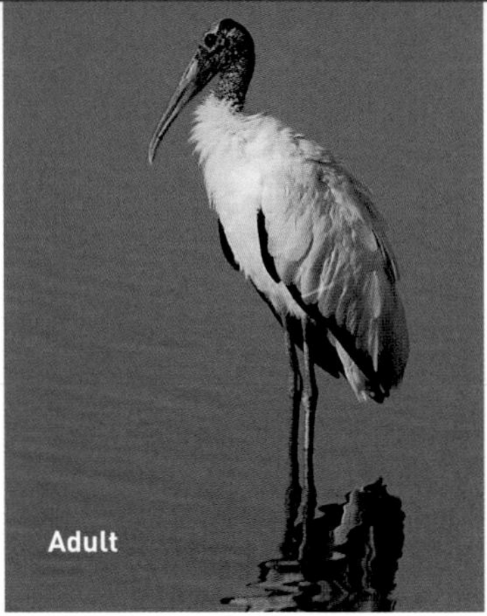

Adult

LOOK FOR: A very large bird with a white body and wings that are starkly black and white. The large down-curved bill and gnarled, featherless head make the Wood Stork a candidate for "ugliest bird." Wood Storks are strong, if slow, fliers and are frequently seen soaring high overhead.

LISTEN FOR: Mostly silent except for some hissing and bill clattering in nesting colonies.

REMEMBER: The Wood Stork is much larger than the White Ibis; has a dark head, unlike the white egrets; and wades instead of swims, like the much larger White Pelican does.

WOW!

The Wood Stork is North America's only native stork, but it's not the stork of the baby-delivery myth—that's the European Stork, which nests on rooftops and chimneys throughout Europe.

▼ *A close look at a Wood Stork's bald head hints at this species' close relationship to the Turkey and Black Vultures.*

FIND IT: Commonly found in southern swamps, lagoons, ponds, and roadside ditches. Colonies nest in large stands of trees, such as cypress. Populations are declining throughout the Southeast because of habitat destruction and alteration.

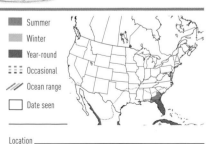

Summer

Winter

Year-round

Occasional

Ocean range

Date seen

Location ______________________

LOOK FOR: A tall, completely gray bird with a red patch on the head and a white patch on each cheek. In flight, Sandhills show both a long neck and long legs. Young Sandhills lack the red crown, and in summer many adults show a lot of rusty brown body feathers.

LISTEN FOR: A low, rattling *graaaahk!* Often compared to the rattle of an American Crow, but louder. This call from flocks high overhead may be your first clue to the presence of Sandhill Cranes.

REMEMBER: The Sandhill Crane differs from the Great Blue Heron in three obvious ways: the Sandhill has a large bustle of feathers on the tail, it has a much stockier body, and it flies with the neck extended, not folded in.

WOW!

More than a quarter million Sandhill Cranes stop over during spring migration along the Platte River in Nebraska. Imagine the noise made by this many large birds as they all take flight at dawn!

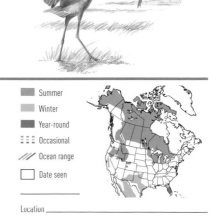

▶ *With loud calls and springing leaps, a courting pair of Sandhill Cranes prepares to mate. Cranes mate for life.*

FIND IT: Widespread and common, especially during spring and fall migration, when large flocks may be seen overhead or foraging in fields, wet prairies, and large marshes.

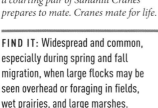

Summer
Winter
Year-round
Occasional
Ocean range
Date seen ____________

Location ____________

Cathartes aura Length: 26"

LOOK FOR: The Turkey Vulture soars with its wings held slightly above horizontal (called a slight dihedral). From below, the Turkey Vulture (TV) shows two-tone wings—black in front, silver in back. The flight silhouette combines long wings and tail with a tiny head (hawks and eagles appear much larger headed in flight).

WOW!

One of the Turkey Vulture's defenses is to puke on an intruder. Trust me: You do not want to get vulture puke on your clothes—you cannot get the smell out!

LISTEN FOR: Normally silent, but will hiss and groan when danger approaches, especially near the nest.

REMEMBER: Turkey Vultures *rock!* When they fly, TVs rock or teeter back and forth, capturing every bit of rising air. So if you see a large raptor rocking in the sky, chances are it's a Turkey Vulture.

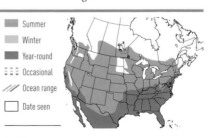

▲ *A Turkey Vulture prepares to deliver a load of predigested roadkill to its nestlings in a hollow log.*

FIND IT: Look up in the afternoon sky in summer and you'll likely see a Turkey Vulture. They are found in open or semi-open habitats but will even venture into heavily wooded areas for their food, which is dead animal flesh, or carrion.

Summer

Winter

Year-round

Occasional

Ocean range

Date seen ______

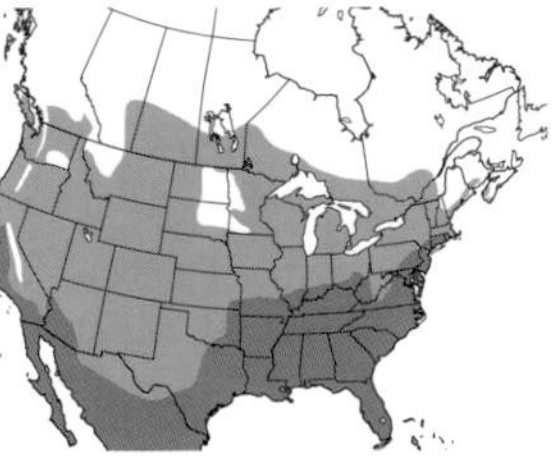

Location ______

LOOK FOR: The Black Vulture is the smaller of our two vultures and in flight looks much more compact and shorter tailed than the Turkey Vulture. The Black Vulture has a featherless gray head (Turkey Vultures have red heads) and when viewed from below shows all-black wings with white tips (primaries).

WOW!
Where both species occur, Black Vultures, although smaller, often bully the Turkey Vultures and drive them away from a feeding site.

LISTEN FOR: Black Vultures are usually silent except near the nest or when threatened, when they issue guttural hisses and barks.

REMEMBER: Black Vultures fly like they are worried they are going to fall out of the sky: lots of fast flapping in between short glides. They hold their wings flat, not slightly raised as Turkey Vultures do.

▶ *A Black Vulture tends its two chicks at a nest inside a decaying tree stump.*

FIND IT: Common in open country and coastal lowlands, the Black Vulture feeds at dumps and even at open dumpsters in cities. Will roost with Turkey Vultures on large power-line towers and in dead trees.

Summer
Winter
Year-round
Occasional
Ocean range
Date seen ____________

Location ____________

Haliaeetus leucocephalus Length: 31"

Adult

Immature (left), adult (right)

WOW!
Now that harmful pesticides such as DDT have been banned, Bald Eagle populations are rebounding. Eagles in the wild are living longer, healthier lives, some as long as 30 years.

LOOK FOR: It's hard to mistake a full-adult Bald Eagle for any other bird, with its huge white (not bald) head, large yellow bill, and white tail. Bald Eagles soar with their wings held flat. They do not rock, or teeter, like Turkey Vultures do.

LISTEN FOR: For such a majestic bird, the Bald Eagle has a wimpy voice. It utters a variety of high-pitched, chattering whistles that sound more like an excited puppy than a national symbol.

REMEMBER: It takes four years for a Bald Eagle to reach full-adult plumage. Until then, it wears a very splotchy pattern of brown and white. In flight, though, an eagle of any age will show the massive wings and huge head that help this species to stand out.

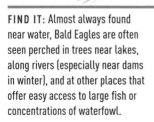

◀ *Bald Eagles don't dive like Ospreys, preferring instead to pick fish off the water's surface. This is an immature Bald Eagle.*

FIND IT: Almost always found near water, Bald Eagles are often seen perched in trees near lakes, along rivers (especially near dams in winter), and at other places that offer easy access to large fish or concentrations of waterfowl.

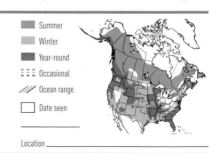

Summer
Winter
Year-round
Occasional
Ocean range
Date seen ______

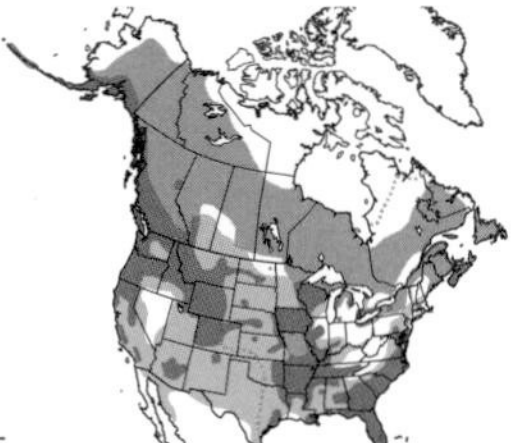

Location ______

LOOK FOR: This magnificent bird soars over the vast open areas of the West on wide dark wings held flat. Adults are almost completely dark brown. A golden hue on the hind-neck feathers gives this bird its name. Juveniles have a bold white base to the tail and white "silver dollar" patches in the middle of their wings.

WOW!

Native Americans greatly admired the Golden Eagle for its powers of flight and its hunting prowess. Eagle feathers are sacred symbols to many native peoples.

LISTEN FOR: Rarely vocalizes, though near the nest it will give barking and whistling sounds.

REMEMBER: Young Golden Eagles have well-defined white patches in the wings and tail. Young Bald Eagles look messier—their white is splotched throughout their plumage until about their third year.

◀ *Golden Eagles are often seen perched along highways waiting for an unwary rabbit. They will kill prey as large as Sandhill Cranes and small deer.*

FIND IT: More common in the West, where it is found from the tundra of the far North, throughout the Great Plains, and south to the deserts and mountains of the Southwest. Always prefers open habitat. In winter they range far and wide in search of food.

Summer

Winter

Year-round

Occasional

Ocean range

Date seen

____

Location ______________

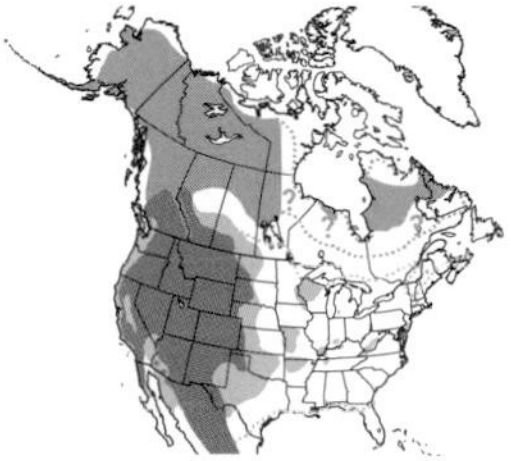

LOOK FOR: Known as the Fish Hawk or Fish Eagle for its preferred food, the Osprey flies with its wings held flat but bent backward at the "wrists." The dark back, white head, and a black raccoon mask make it resemble a Bald Eagle, but the Osprey always has a white breast and belly. The female often shows a dark "bra" band across the chest.

LISTEN FOR: Ospreys are very vocal, giving a loud, clear, high-pitched whistle—*tyou-tyou-tyou-tyou*—that sounds like they are giggling.

REMEMBER: A soaring Osprey appears crooked winged, more like a gull than a raptor from a distance. Eagles and most other raptors rarely show bent-back wings.

▼ *Young Ospreys remain on the tropical wintering grounds through their second summer, honing their fishing skills.*

WOW!

Bald Eagles often let the Osprey do the hard work of catching a fish; then they'll chase after the Osprey to steal its food—not very nice behavior from our national symbol.

FIND IT: Common near large bodies of water, the Osprey soars over the water looking for fish. When it sees one, the Osprey dives talons-first into the water, then carries the fish to a perch where it can carve up its victim.

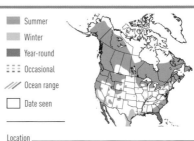

Summer

Winter

Year-round

Occasional

Ocean range

Date seen

____

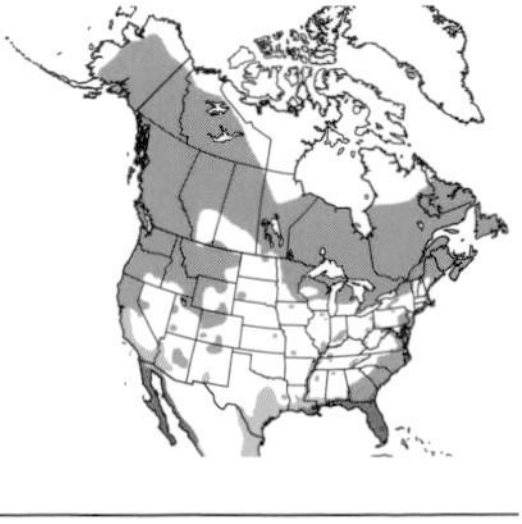

Location ______________

LOOK FOR: The Mississippi Kite is the smallest of our four kite species and also the most widespread. In flight, the adult shows an all-black tail, and the male shows white secondaries on the upperwing. Juvenile birds are streaky dark brown overall.

LISTEN FOR: A very high-pitched, whistled *fee-feww!* that descends in tone. It sounds similar to a Broad-winged Hawk's call. A second call is a more percussive (and very shorebirdlike) *fee-tititititi!*

REMEMBER: A flying Mississippi Kite can look a lot like a Peregrine Falcon. The kite's wings and body are slimmer than the falcon's. Its flight is light and buoyant compared to the powerful, direct flight of a Peregrine.

▲ *Mississippi Kites catch dragonflies on the wing and devour them in flight.*

WOW!

Mississippi Kites are such good fliers that they can catch large flying insects and eat them in flight! To accomplish this, a kite bends its head down and reaches its feet upward.

FIND IT: Most common in the Southeast from spring through fall, where it soars high over woodland edges and swamps. Each spring and fall, a few Mississippi Kites are found far outside their normal range. They winter in South America.

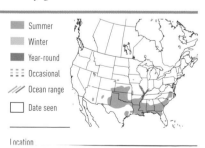

Summer
Winter
Year-round
Occasional
Ocean range
Date seen

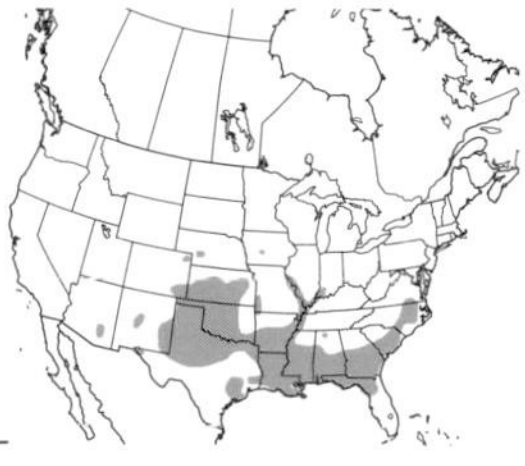

Location

SHARP-SHINNED HAWK

Accipiter striatus Length: 13"

LOOK FOR: The Sharp-shinned Hawk is the smaller of our two common accipiters—the bird-chasing hawks. Adults have blue-gray backs and orange-spotted breasts. Young birds are brown overall with large brown blotches on white breasts (young Cooper's Hawks have finer breast streaks). In flight, the Sharpie's head does not extend much past the front of the wings, but the Coop's does.

LISTEN FOR: Not a very vocal raptor. The Sharpie gives a high, rapid-fire *tyoo-tyoo-tyoo-tyoo* call, often near the nest.

REMEMBER: In general, Sharpies look smaller headed and thinner legged (when perched) than Cooper's Hawks. They also flap faster when flying.

WOW!

Sharp-shinned and Cooper's Hawks are built with compact wings for short, speedy bursts of flight and for maneuvering through thick woods in pursuit of their main prey: other birds.

▸ *A Sharp-shinned Hawk being mobbed by Blue Jays.*

FIND IT: You are likely to see a Sharpie rocketing through a woodland clearing scattering panicked song-birds, perched along a wooded edge watching for prey, or soaring in tight circles overhead.

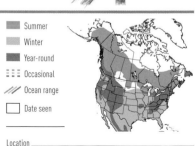

Summer

Winter

Year-round

Occasional

Ocean range

☐ Date seen ______________

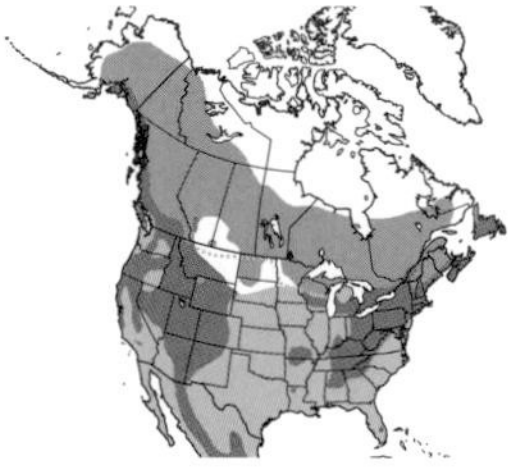

Location ______________________________

COOPER'S HAWK

Accipiter cooperii Length: 19" (female) to 14" (male)

LOOK FOR: The Cooper's Hawk (or Coop) is chunkier than the Sharp-shinned. Bigger headed with thicker legs and a longer tail that often shows a rounded tip (Sharpie tails can appear squared off), the Coop frequently chases and kills larger prey, including Blue Jays, Mourning Doves, and Northern Flickers! Adults have blue-gray backs and reddish orange chests. Juveniles are streaky brown in front with brown backs.

WOW!

To keep the element of surprise on their side, Coops and Sharpies will spiral high to dive on unsuspecting birds. The birds below cannot see the predator coming out of the sun.

LISTEN FOR: Not very vocal except when disturbed near the nest; then it utters a flickerlike *ka-ka-ka-ka*.

REMEMBER: In general, Coops fly with slower, stronger flaps than Sharpies, and their heads stick out farther in front of their wings.

▶ *A Cooper's Hawk hunts at a backyard feeding station, spooking a Tufted Titmouse and a male Purple Finch.*

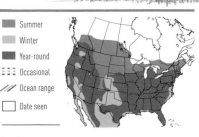

FIND IT: Increasingly common in suburban and urban settings, Coops may be more likely to perch out in the open than Sharpies. Look for very thick-looking legs. When you see songbirds scatter suddenly, look for an accipiter.

Summer
Winter
Year-round
Occasional
Ocean range
Date seen

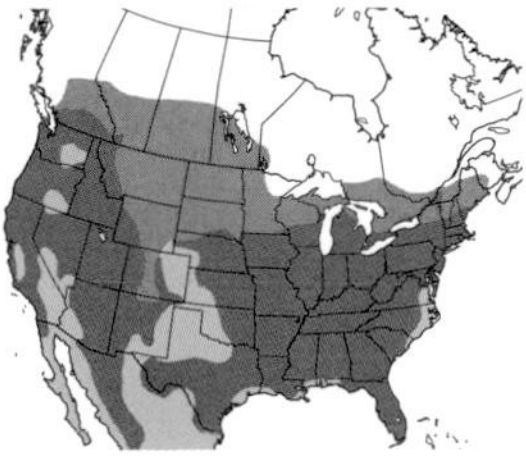

Location ______

Circus cyaneus Length: 18"

Adult male

Adult female

LOOK FOR: The Northern Harrier is a long-winged, long-tailed, and lanky hawk most often seen gliding low over marshy grasslands, hunting for small mammals and birds. The best field mark is its white rump patch, which harriers have in all plumages. Adult females and juvenile birds are warm brown overall (young birds are almost orange-brown).

LISTEN FOR: A chattering, squeaky *chew-chew-chew-chew-chew*, given in a long series. Also gives a thin, high-pitched whistle that slurs downward in pitch, *tssieww!*

REMEMBER: Harriers can change their flight shape. Always look for the long-winged profile and for that distinctive white rump.

WOW!

The feathers that outline its face help to channel sounds to the harrier's ears, much like the feathered faces of several owl species. The harrier is built to hunt by both sight and sound.

▶ *The Northern Harrier's round facial disk of feathers give it an owl-like appearance.*

FIND IT: Though formerly called Marsh Hawk, Northern Harriers can be found in a variety of open habitats—including along beaches, airport runways, and dry prairies. They often perch on the ground or on low fenceposts.

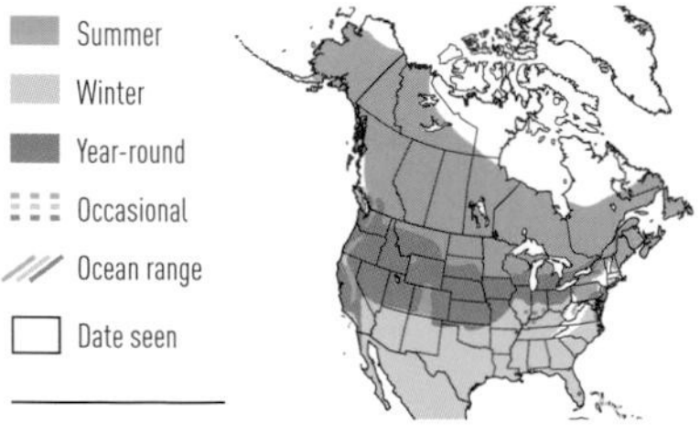

Location ______________________

LOOK FOR: A stunning bird of the desert Southwest, the Harris's Hawk is mostly dark tones of rufous (shoulders and legs) and chocolate (body and head). The dark tail has a broad white base and white tip. Bill has a noticeable yellow base, or cere. Legs are long and yellow.

LISTEN FOR: A harsh, loud *raaaack!*

REMEMBER: The all-dark body, rufous legs, and bold white tail base are field marks unique to this species.

WOW!

Two male Harris's Hawks may mate with one female, and all three will work to raise the young.

▶ *Family groups of Harris's Hawks hunt cooperatively. Some birds flush the prey while others pursue and kill it.*

FIND IT: Uncommon in mesquite woods and thorny desert habitat. Also found along wooded rivers. Often seen perched in small groups of three or more birds.

Summer
Winter
Year-round
Occasional
Ocean range
Date seen

____

Location ______________

RED-SHOULDERED HAWK

Buteo lineatus Length: 17"

LOOK FOR: Smaller than a Red-tailed but larger than a Broad-winged, the Red-shouldered Hawk has a chunky body and long wings. Best field mark for this bird in flight is the pale crescent near each wingtip, called *wing windows* by birders.

LISTEN FOR: One of our most vocal raptors, the Red-shouldered Hawk gives a high scream: *keeyah!* This call is repeated often, especially by flying birds during spring and summer. Blue Jays may imitate this call, even using it to scare other birds away from bird feeders.

REMEMBER: Red-shoulders flap their wings quickly two or three times in between glides, much like a Cooper's Hawk does.

WOW!

Resident Red-shoulders will harass migrant hawks that fly over their nesting territories, calling loudly and dive-bombing other buteos and even eagles!

▶ *Red-shouldered Hawks perch at the edge of the woods, watching for the movement of prey on the ground.*

FIND IT: Common in southeastern woodlands, less common in the North, this hawk is often heard before it is seen. Prefers wooded habitat near water in spring and summer, and woodland edges in winter.

Summer
Winter
Year-round
Occasional
Ocean range
Date seen ______

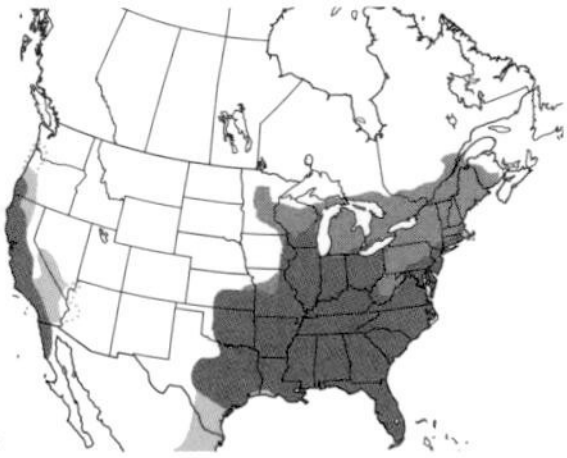

Location ______

Buteo platypterus Length: 15"

LOOK FOR: The Broad-winged Hawk is our smallest buteo (it's smaller than an American Crow) and has a very compact shape in flight. From below, it appears very pale, with a black outline on wings and contrasting black-and-white bands on the tail. Adults have rusty chests and "wingpits." Young birds often lack the distinctive chest and tail markings of adults.

LISTEN FOR: In spring, listen for the Broad-winged's high-pitched, two-note whistle: *tee-teeeee!*

REMEMBER: Soaring hawks can be confusing, but if your bird has broad black-and-white tail bands and a black outline on white wings, you've got a Broad-winged Hawk.

WOW!

More than 400,000 Broad-winged Hawks were counted passing by a fall hawk-watching station in a single day in Veracruz, Mexico!

► *Riding rising areas of hot air, known as thermals, a kettle of Broad-winged Hawks gains altitude in fall migration.*

FIND IT: In fall, Broad-winged Hawks gather into large flocks as they head to South America for the winter. Scan the sky on warm, sunny days between mid-August and mid-October for their soaring flocks, called *kettles*.

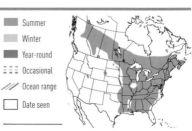

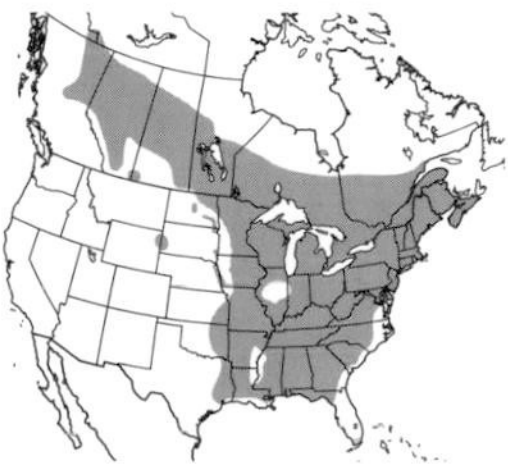

____

Location __________

Buteo jamaicensis Length: 19"

LOOK FOR: A large dark-backed raptor. The adult's rust-colored tail (which earned this species its name) is most obvious on soaring birds. On perched birds the dark belly band and light-colored chest often stand out, and they often show a white V across the upper back. Red-tails may soar in large smooth circles or they may hover in place, scanning for prey.

WOW!

The Red-tailed Hawk's wild-sounding call is often used in place of the Bald Eagle's call on TV and in movies.

LISTEN FOR: A high-pitched scream that sounds like steam escaping from a pipe—*fscheeeew!*—is most often uttered by flying birds.

REMEMBER: Red-tails are the most widespread of our large soaring hawks. Their coloration can vary from very light to very dark, so overall size and shape is a good starting point to ID them.

▶ *Most perched Red-tailed Hawks show the white V of backpack straps across their backs.*

FIND IT: Look for Red-tails perched or soaring where they can command a clear view of open habitat where small mammals live. In late winter, look for pairs perched close together. The larger one is the female.

Summer
Winter
Year-round
Occasional
Ocean range
Date seen

Location ____________

SWAINSON'S HAWK

Buteo swainsoni Length: 19"

LOOK FOR: Compared to other buteos, such as the Red-tailed Hawk, the Swainson's Hawk has a thinner body and longer, more tapered wings in flight. The white inner linings on the wings (in flight) are a key field mark, as is the dark brown bib above a white breast. Soaring hawks hold their wings above horizontal, similar to a Turkey Vulture, but do not rock like a TV.

LISTEN FOR: The Swainson's raspy, descending cry—*cree-yaah*—sounds like a cross between a Red-tailed Hawk's scream and a cat's *meow*.

REMEMBER: Swainson's Hawks come in both light and dark color variations (morphs), so base identification on overall size and shape in flight.

◄ *Swainson's Hawks are primarily insect eaters. Grasshoppers are a favorite food.*

WOW!

Swainson's Hawks sometimes migrate in huge flocks. One wintering flock in Argentina contained more than 12,000 birds!

FIND IT: Swainson's Hawks spend the winter eating insects in South America, and in spring they head back to the Great Plains, where they nest in isolated groves of large trees.

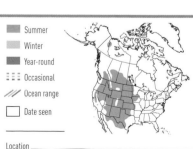

Summer

Winter

Year-round

Occasional

Ocean range

☐ Date seen

____

Location ______________________________

FERRUGINOUS HAWK

Buteo regalis Length: 23–24"

LOOK FOR: A large, pale soaring hawk with long narrow-tipped wings. Dark brown leg feathers form a noticeable dark V on birds viewed overhead. Flight feathers (primaries and secondaries) always appear clean and white. Perched birds appear rufous backed and pale breasted.

LISTEN FOR: Call is a soft, whistled *keee-www*.

REMEMBER: The wings of the Ferruginous Hawk are longer and more tapered than those of the Red-tailed and Swainson's Hawks. This gives this species a "wingier" look than other buteos have.

WOW!

When bison roamed the plains, some Ferruginous Hawks built their nests from bison bones and lined them with bison "patties."

◀ *Ferruginous Hawks may wait on the ground for an unwary mammal or snake to come out of its burrow.*

FIND IT: Uncommon across the dry, open plains and prairies of the West, where it hunts for small mammals, birds, and snakes.

Summer

Winter

Year-round

Occasional

Ocean range

☐ Date seen

____

Location __________

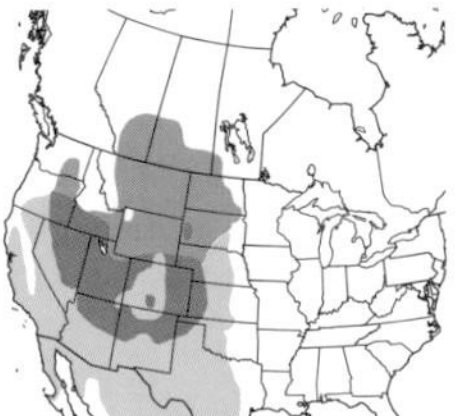

PEREGRINE FALCON

Falco peregrinus Length: 17" Wingspan: 41"

LOOK FOR: Large and powerful, the Peregrine looks like a kestrel on steroids. It flies with smooth but shallow wing strokes, easily generating great speed. Overall it appears dark. Young birds are brown overall, and adults have gray backs and white breasts barred with black. All Peregrines have black mustaches, though they can be hard to see on distant birds.

LISTEN FOR: Normally silent but near nest will utter a loud, raspy *klee-klee-klee-klee!*

REMEMBER: When a Peregrine flies into sight, waterfowl and shorebirds usually take off in a panic. This is a good clue to the presence of this powerful hunter, so scan the skies for a Peregrine whenever you see fleeing birds.

▲ *A Peregrine Falcon pair courting high above a city, where they will nest on a building ledge.*

WOW!

Peregrine Falcons may reach speeds of 200 miles per hour when diving for prey. They use their balled-up talons to knock out their prey, then catch the hapless, falling bird before it hits the ground or water.

FIND IT: Found worldwide but not common anywhere, Peregrines like to perch up high (on towers, cliffs, buildings, bridges), where they can watch for prey.

Summer
Winter
Year-round
Occasional
Ocean range

☐ Date seen ______________

Location ______________

MERLIN

Falco columbarius Length: 11–12"

LOOK FOR: This small dark falcon is slightly larger than the more common American Kestrel. A perched Merlin looks darker and stockier than a kestrel. There are several races of Merlin across North America, but all have a uniformly dark back and a banded tail.

LISTEN FOR: Call is a high-pitched *kee-kee-kee-kee.*

REMEMBER: Merlins usually fly purposefully—as if they're late for something. They rarely hover in one place, like kestrels do.

WOW!

Perch-and-wait hunters, Merlins watch for a passing bird and then pursue and capture their prey in flight.

▶ *Merlins are increasingly nesting in urban settings across the Midwest, living on a diet of House Sparrows.*

FIND IT: Prefers open habitat throughout its range. Open conifer forests, prairie grassland with scattered trees, coastal marshes. Can be seen anywhere in migration.

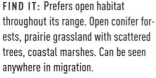

Summer

Winter

Year-round

Occasional

Ocean range

Date seen

____

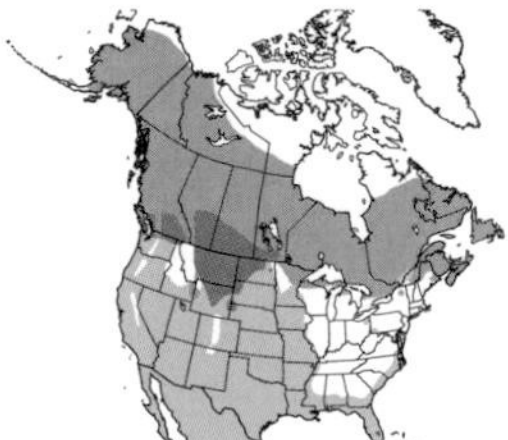

Location ______________________________

AMERICAN KESTREL

Falco sparverius Length: 10" Wingspan: 22"

Adult male

Adult female

LOOK FOR: Our smallest falcon, the American Kestrel is a familiar sight across North America, hovering over meadows or sitting on a power line, flicking its tail backward. Adult females are mostly streaky brown overall, with black stripes on rusty backs. Adult males are more colorful, with bluish wings and headbands and rusty chests, tails, and backs. Both sexes have black mustache stripes.

LISTEN FOR: The American Kestrel's primary call is a high-pitched, excited-sounding *killy-killy-killy-killy.*

REMEMBER: In flight, falcons usually show pointed wings. Kestrels appear lighter and "bouncier" than other, larger falcons.

WOW!

The spots on the back and sides of a kestrel's head look like a face. These markings are meant to fool predators into thinking the kestrel is looking at them and is prepared for an attack.

▶ *False eyespots on the kestrel's nape make it appear to be looking at you with eyes in the back of its head.*

FIND IT: Kestrels love open grassy areas where they can hunt for small mammals and large insects. Watch for them perched along fence lines and power lines and in the tops of trees. A small falcon hovering in place is almost surely a kestrel.

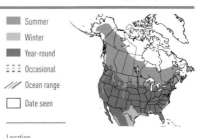

Summer

Winter

Year-round

Occasional

Ocean range

☐ Date seen

____

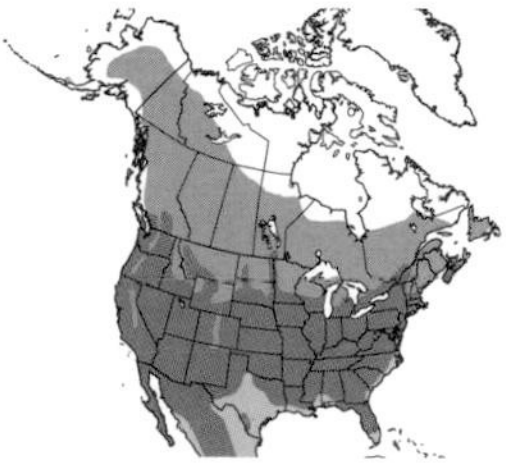

Location ___________

Fulica americana Length: 15"

LOOK FOR: The American Coot is a sooty gray football-shaped chunk of a bird with a stubby white bill and a white-and-rust forehead patch. Although the coot looks and acts like a duck, it's actually a member of the rail family.

LISTEN FOR: Coots are noisy birds, uttering a variety of grunts, cackles, chatters, and croaks.

REMEMBER: Coots bob their heads forward and back when swimming. In order to fly, coots have to scamper across the surface of the water to gain enough speed to become airborne.

WOW!

Coot hatchlings can swim very well within a few hours of hatching. Coots are aggressive birds that often steal food items from nearby ducks. Bullies!

▼ *American Coot chicks look like bald men in turtleneck sweaters. Their bright heads stimulate the parents to feed them.*

FIND IT: Widespread on lakes, ponds, and marshes all across North America, coots are as at home walking on land as they are swimming on water. Suburban ponds and golf courses seem to be especially attractive to coots.

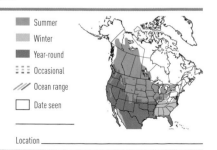

Summer
Winter
Year-round
Occasional
Ocean range
Date seen ________

Location ____________________

Gallinula chloropus Length: 14"

LOOK FOR: A charcoal gray bird with a rusty back, white body striping, and a red bill, the Common Moorhen is equally at home swimming in water and walking on land. Both sexes have the bright red bill and forehead shield, though the male's is brighter.

LISTEN FOR: Whinnying, squeaky notes in a series, slowing down near the end. Sounds like someone is torturing a frog. Also, a high, sharp *peek!*

REMEMBER: Distant moorhens can be told apart from coots by the white horizontal side stripes and extensive white on the tails.

WOW!

Q: What looks like a duck but is not a duck? A: A Common Moorhen, which is more closely related to cranes and rails than to ducks (even though they swim like ducks).

◀ *How a Common Moorhen swims so well without webbed feet is a mystery. Perhaps its bobbing head helps it motor along.*

FIND IT: You may hear a moorhen before you see it, but look for it in or near the dense vegetation along the edge of freshwater marshes and ponds doing a "funky chicken" motion as it swims or walks.

Summer
Winter
Year-round
Occasional
Ocean range
Date seen ____________

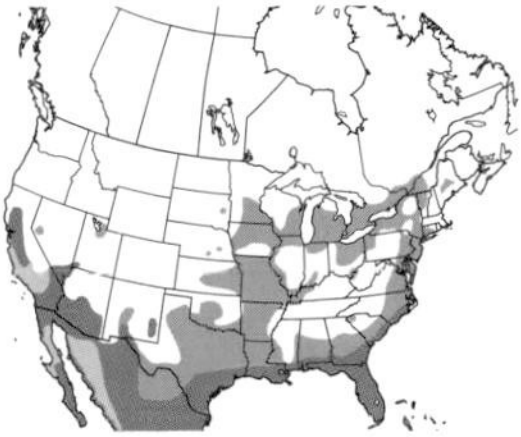

Location ____________

PURPLE GALLINULE

Porphyrio martinica Length: 14"

LOOK FOR: While not very widely distributed, the Purple Gallinule is likely to make you say, "Wow!" Its football-sized body is decked out in hues of purple, blue, and green, and its legs and bill tip are bright yellow. A red stripe covers the base of the bill, and a light blue shield covers the forehead.

LISTEN FOR: A series of high-pitched peeps and nasal clucks that sound like a maniacal chicken.

REMEMBER: The Purple Gallinule is like a Technicolor version of the Common Moorhen. From a distance, the gallinule looks all dark while the moorhen shows white in the wings. Seeing the purple-green body should be fairly easy except in poor light.

WOW!
Another name for the Purple Gallinule is Swamphen.

◀ *The Purple Gallinule's long yellow toes allow it to step lightly over the lily pads without sinking or swimming.*

FIND IT: The Purple Gallinule's long narrow toes, when spread out, distribute the bird's weight, enabling it to walk on lily pads and other floating vegetation in wetlands and freshwater marshes.

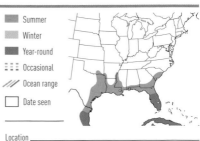

Summer
Winter
Year-round
Occasional
Ocean range
Date seen

Location ______

SEMIPALMATED PLOVER

Charadrius semipalmatus Length: 7¼"

LOOK FOR: Among our small plovers, the Semipalmated Plover is the most frequently encountered, especially in spring and fall migration. It looks like a miniature Killdeer but has only one dark breast-band (the Killdeer has two). The Semipal's dark back helps it to blend into the wet mud flats it prefers.

LISTEN FOR: The Semipalmated Plover gives three different calls. In flight, it gives a rising, throaty *too-wee.* Threatened birds give a rapid *doi-doi-doi-doi-doi* and a dry, burbly *dwiip.*

REMEMBER: The Killdeer is a full three inches larger than a Semipalmated Plover. Other small plovers are less common inland.

WOW!

In the 1800s, Semipalmated Plover populations were decimated because the birds were hunted for the plume trade. Imagine shooting something as small and cute as a Semipal Plover!

▶ *Semipalmated Plovers are commonly seen on mud flats pulling marine worms, like Robins on a lawn.*

FIND IT: Though they nest far to the north, Semipalmated Plovers are commonly seen in muddy wetlands, on mud flats, and on beaches in migration, where they scamper along plucking food items from the mud's surface.

Summer
Winter
Year-round
Occasional
Ocean range
Date seen

____

Location ______________

Adult, breeding

LOOK FOR: A small, pale, and stocky plover, the Piping Plover has yellow legs and a plain face that help distinguish it from its more common relatives. In spring and summer, the adult has an orange bill with a black tip and a single narrow breast-band that may or may not connect in front.

WOW!

Recovery efforts are helping to protect Piping Plover nesting areas on heavily used beaches along the Atlantic Coast. If you visit there, be sure to obey the signs and avoid disturbing these birds.

LISTEN FOR: This bird could be called the Peeping Plover for the variety of mellow, whistling *peeps* it utters.

REMEMBER: The closely related Semipalmated Plover is much more widespread and common than the Piping Plover. Semipals have dark brown backs and boldly marked heads and faces. Piping Plovers are pale.

◀ *Piping Plover chicks are able to run and pick up food on their first day out of the egg. The parents simply supervise them.*

FIND IT: Piping Plovers nest on dry sandy beaches along the Atlantic Coast and on dry flats near lakes and rivers in the upper Great Plains. Look for them on dry pale ground that matches their back color.

Summer
Winter
Year-round
Occasional
Ocean range
Date seen ________

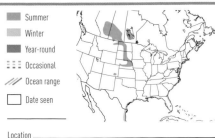

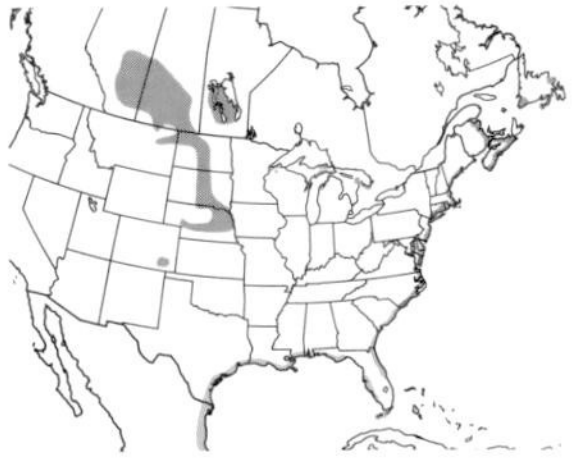

Location ________

LOOK FOR: A large, noisy shorebird that is often found far from the shore, the Killdeer has two black breast-bands on a white chest. In flight, the Killdeer looks long and slender and shows a bright orange rump and black wingtips and tail tip.

LISTEN FOR: The Killdeer is said to call its name as it flies, but really it sounds more like *tee-deee, tee-deee*. High-pitched and loud, the Killdeer's call is hard not to notice (that's why its Latin name means "vociferous").

REMEMBER: No other commonly encountered shorebird has the Killdeer's double black breast-bands.

▼ *A Killdeer distracts attention from its well-camouflaged eggs. It flashes its rump, flutters, and calls.*

WOW!

Killdeer perform a distraction display, to lure predators away from their nest, flopping around on the ground and faking a broken wing.

FIND IT: The widespread and common Killdeer can be found on open grassy meadows, on ball fields, along gravel roads, at airports, and on mud flats or beaches. Look for them running, foraging, and calling to one another.

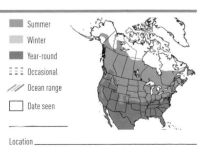

Summer
Winter
Year-round
Occasional
Ocean range
Date seen

Location ____________

AMERICAN OYSTERCATCHER

Haematopus palliates Length: 18"

LOOK FOR: The American Oystercatcher is a large, stocky, boldly marked shorebird with a big, straight orange-red bill jutting out from a black head. In flight, the oystercatcher looks white below and black above and seems to flash black and white as it flies. Adult males and females are identical.

LISTEN FOR: Long call is a series of loud, high-pitched *whee-eeps* followed by a rapid *didididididididi* dropping in tone and slowing. Short call is a loud *peep* or *weep!*

REMEMBER: You might confuse the oystercatcher in flight with a Willet—both flash a lot of white in the wings. But the oystercatcher's large orange-red bill should stand out, even from a distance.

WOW!

The oystercatcher uses its powerful bill to open the shells of oysters, clams, and mussels. If you've ever tried to do this, you know how hard it is. Imagine doing it without hands!

◀ *An American Oystercatcher pries a limpet away from a rock as its chick looks on.*

FIND IT: Always found near salt water, the American Oystercatcher is a resident along the Atlantic and Gulf coasts, where it frequents beaches, mud flats, marshes, sandbars, and islands.

Summer
Winter
Year-round
Occasional
Ocean range
Date seen ______

Location ______

LOOK FOR: Named for its long slender legs, the Black-necked Stilt is a delicate-looking bird with a needlelike bill and slender overall appearance. Black above, white below, with bright pinkish red legs.

LISTEN FOR: Loud, frantic-sounding *kleep-kleep-kleep!* Often given when disturbed.

REMEMBER: The stilt has a straight bill. The winter-plumaged American Avocet looks similar but is larger and has an upturned bill.

WOW!

Though they look delicate, Black-necked Stilts are tough birds. They spend most of their lives, including nesting, on hot, sun-baked ground, such as salt flats.

◀ *Stilts use their thin, needlelike bills to pick up their prey.*

FIND IT: Common in shallow ponds, marshes, and mud flats, where it walks delicately, picking up small insects and crustaceans. Often in mixed flocks with American Avocets. Most stilts winter in coastal habitat.

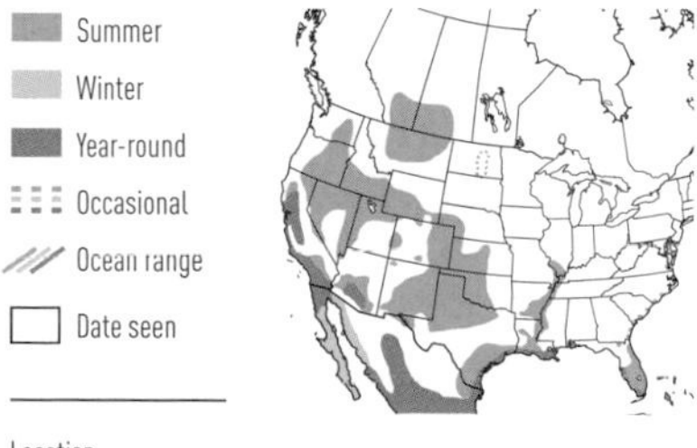

Date seen ____________

Location ____________

AMERICAN AVOCET

Recurvirostra americana Length: 18"

LOOK FOR: The graceful curves of the American Avocet's bill and neck serve it well as it feeds while walking along, sweeping the bill back and forth in shallow water. In spring and summer, adults have rusty heads and necks. In winter, the heads are gray. Black wings are divided by white.

LISTEN FOR: A clear, high-pitched *pwee-eep!*

REMEMBER: In winter plumage, the American Avocet's head is gray, making it look more like the Black-necked Stilt than it does in summer, when its head is rusty. Check the bills and legs. Avocets have long blue-gray legs and upturned bills. Stilts have pink legs and straight bills.

WOW!

The bill of the female avocet is more sharply upturned than that of the male. Why? No one knows.

▶ *An American Avocet calls an alarm when danger threatens the nest.*

FIND IT: Avocets are found in shallow marshes, lakes, and wetlands. They nest in loose colonies. They are often found in flocks, sometimes associated with Black-necked Stilts.

Summer
Winter
Year-round
Occasional
Ocean range
Date seen ____

Location ____

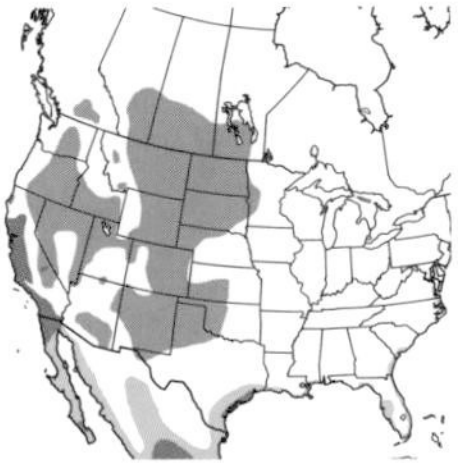

Tringa semipalmata Length: 15"

WOW!
Three weeks after young Willets hatch, the mother abandons them to the care of the dad, who takes care of the young for several weeks until they are able to fend for themselves.

LOOK FOR: In all plumages, the Willet is nothing special to look at until it flies and flashes its boldly patterned black-and-white wings. A large, chunky shorebird with a straight, stout bill, the Willet changes its mottled brown breeding plumage for conservative gray in winter.

LISTEN FOR: A loud ringing call in which the Willet says its name: *pill-will-willet!*

REMEMBER: Compared to other common shorebirds, Willets appear very plain overall but with a solid build—like a shorebird that's been working out with weights.

▶ *Willets look like big, generic gray shorebirds until they open their stunning black-and-white wings.*

FIND IT: Very common along the Atlantic and Gulf coasts, where it prefers saltwater marshes, swampy meadows, and beaches. Western birds nest on inland marshes and grasslands near water.

Summer
Winter
Year-round
Occasional
Ocean range
Date seen

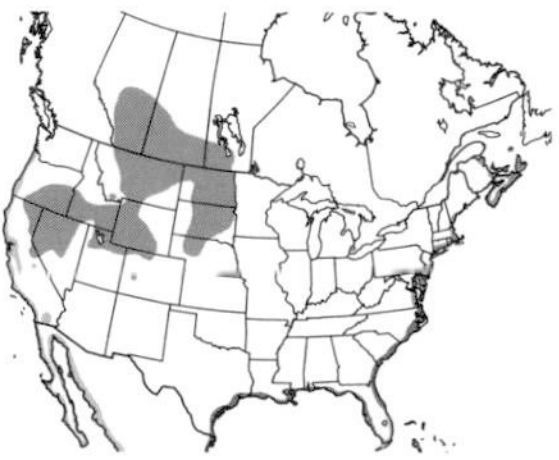

____

Location ______________

GREATER and LESSER YELLOWLEGS

Tringa melanoleuca, Tringa flavipes   Length: 14", 10½"

LOOK FOR: Greaters are larger than Lessers, but size can be hard to judge unless both species are side by side. Greaters also have a longer, thicker bill, especially at the base, that is often two-tone. Lessers appear delicate in every way, including the all-dark needle-thin bill. Both have long, bright yellow legs.

WOW!

Lesser Yellowlegs, despite being smaller birds, are less easily spooked into flight than Greater Yellowlegs. Many birders have noticed this as they've approached flocks of both species of yellowlegs.

LISTEN FOR: Voice is the best way to tell these birds apart. Greaters tend to call in three or four loud, clear notes: *teww-teww-teww-teww!* Lessers call less, that is, their calls are shorter, with one or two mellow notes: *ti-teww, ti-teww.*

REMEMBER: Greaters have longer, thicker bills and longer, louder calls, and you have a *greater* chance of seeing a Greater Yellowlegs in cold weather.

▶ *The yellowlegs' names refer to the relative size difference between the two species.*

FIND IT: During migration both species can be found on marshes (saltwater or freshwater) and mud flats and along streams and rivers. Lessers seem to prefer smaller bodies of water. Greaters remain farther north in winter.

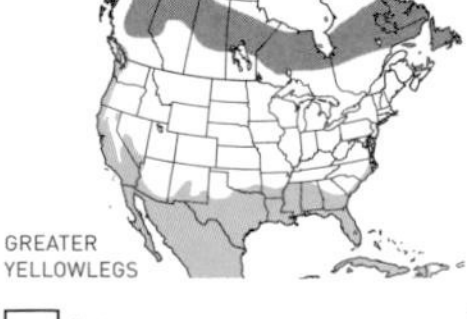

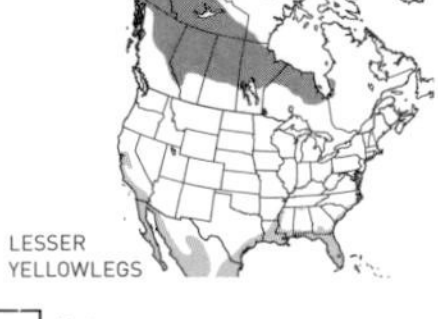

☐ Date seen ______________ ☐ Date seen ______________

Location ______________ ______________

LOOK FOR: A slender shorebird with dark wings and back dotted with tiny white spots, the Solitary Sandpiper is often found alone, as its name suggests. Its white eye-rings give it a spectacled look, as though it were wearing a pair of white glasses.

LISTEN FOR: High-pitched, piercing whistle, *peet-deet, peet-deet-deet,* often given in flight or when alarmed.

REMEMBER: Solitaries are darker overall than Spotted Sandpipers and have longer necks and more distinct white eye-rings than Spotties.

▲ *A Solitary Sandpiper tends its eggs in a twig nest originally built by a Gray Jay.*

WOW!

The Solitary Sandpiper does not nest on the ground like most other shorebirds. It nests in old birds' nests in the tops of trees, as high as 40 feet off the ground!

FIND IT: Solitary Sandpipers prefer the muddy edges of ponds, creeks, and wetland marshes. They often bob their tails in a manner similar to that of the Spotted Sandpiper.

Summer

Winter

Year-round

Occasional

Ocean range

Date seen

____

Location ______________

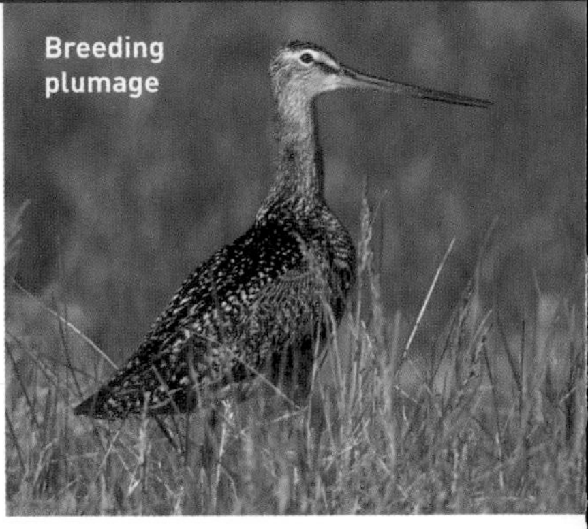

LOOK FOR: The largest of our three godwit species, the Marbled Godwit shows cinnamon underwings in flight. Its overall warm brown appearance in all plumages lacks obvious contrast. The long upcurved bill is pink at the base and black at the tip.

LISTEN FOR: Call is a two-note *god-WIT!* or a harsh single note: *kerr.* Flight display song sounds like *TheRabbitTheRabbit The Rabbit!*

REMEMBER: Similar to Long-billed Curlew but with an upturned bill (the curlew's bill is long and downward-curving).

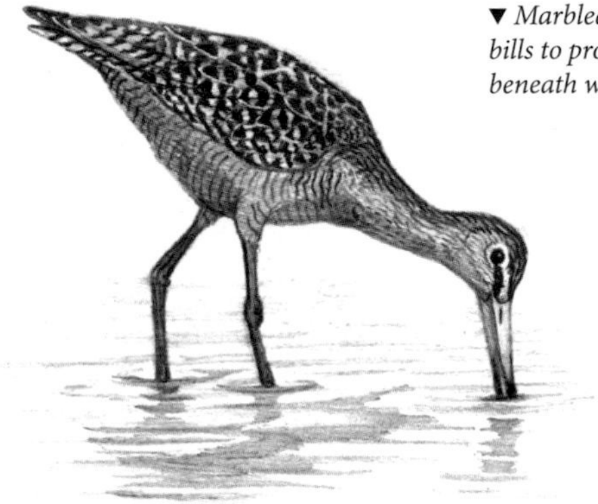

▼ *Marbled Godwits use their long bills to probe for food hiding beneath water or mud.*

WOW!

Probing deep in the ground with their long bills, godwits locate their food by touch. The godwit's diet includes crabs, earthworms, leeches, grasshoppers, and plant tubers.

FIND IT: Breeds in wet meadows in the prairie potholes region of the upper Great Plains. Winters in flocks on beaches, tidal mud flats, and marshes in coastal regions.

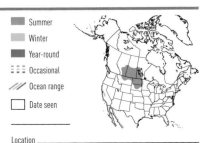

Summer

Winter

Year-round

Occasional

Ocean range

☐ Date seen

____

Location _______________

LOOK FOR: A large gray-brown bird on sturdy gray legs with a down-curved bill. Unlike other large shorebirds, Whimbrels do not show obvious field marks in flight, appearing rather plain overall. The dark stripes on the Whimbrel's head also set it apart from other large, long-billed shorebirds.

LISTEN FOR: A series of six or seven clear, whistled notes on the same pitch: *tu-tu-tu-tu-tu-tu-tu.*

REMEMBER: Flying Whimbrels show no obvious field marks. They are plain Janes.

WOW!

Another name for the Whimbrel is Short-billed Curlew. This name must have been a comparison to the Long-billed Curlew of western North America, which has a really long bill.

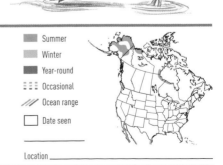

▶ *The Whimbrel's long, curved bill is useful for handling prickly prey such as fiddler crabs.*

FIND IT: Most common on both coasts, where migrant flocks can be found in grassy marshes and on mud flats. Look for Whimbrels as they forage, picking up insects with their long, decurved bills.

Summer
Winter
Year-round
Occasional
Ocean range
Date seen

____

Location ______________

Adult, breeding

Adult, nonbreeding

LOOK FOR: From April through September, the Ruddy Turnstone has a zebra-striped face and breast and rusty (or ruddy) wings. The rest of the year, the bird wears a much-faded version of this plumage. Its short legs remain orange year-round.

LISTEN FOR: A very ternlike *klew!* alarm call, plus an unmusical chatter, *kkkkkkkrrrkkkkkrr.*

REMEMBER: Even in its dull winter plumage, the Ruddy Turnstone retains some of the pattern on its head and breast.

WOW!

The turnstone is named for the way it finds food, using its bill to turn over stones. Its Latin name translates to "sandy place interpreter" for its habit of calling out to warn other birds of danger.

◀ *A Ruddy Turnstone flips over a stone.*

FIND IT: Sandy or rocky beaches, breakwaters, and jetties are the preferred habitat of the Ruddy Turnstone. Most commonly found along coasts in migration and winter, rarer inland.

Summer

Winter

Year-round

Occasional

Ocean range

Date seen ______

Location ______

SANDERLING

Calidris alba Length: 8"

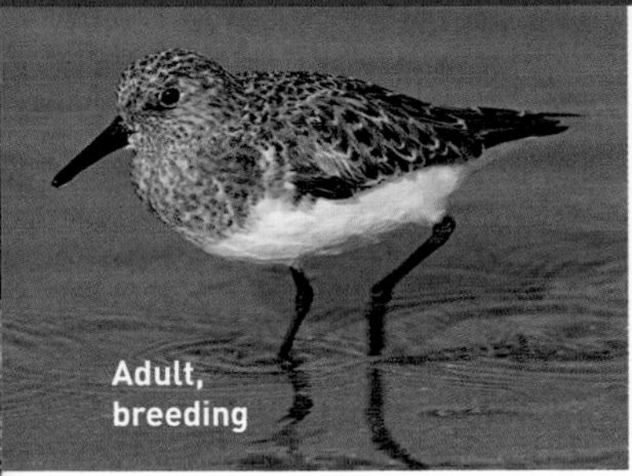

LOOK FOR: Those small, plump, pale shorebirds running back and forth with each wave on the beach are Sanderlings, the stereotypical birds of the seashore. We almost never get to see the Sanderling's rich rufous breeding plumage, worn only from May through August on its Arctic breeding grounds. The rest of the year the Sanderling is pale gray above and bright white below, with black legs and bill.

WOW!

Sanderlings have an uncanny ability to time their feeding dashes between waves. They pursue a wave as it retreats, probe for exposed sand crabs, then run away as the next wave comes in.

LISTEN FOR: A short, sharp *queet* or *queet-queet-queet* often uttered in flight.

REMEMBER: Sanderlings are paler and more active and nimble than most other small sandpipers.

► *Sanderlings probe in the mud for food. They always seem to be in a great hurry; perhaps it's because their prey moves so fast.*

FIND IT: On sandy beaches along oceans and lakes, Sanderlings will be found dashing to and fro like wind-up toys. Often found in large flocks during migration and in winter.

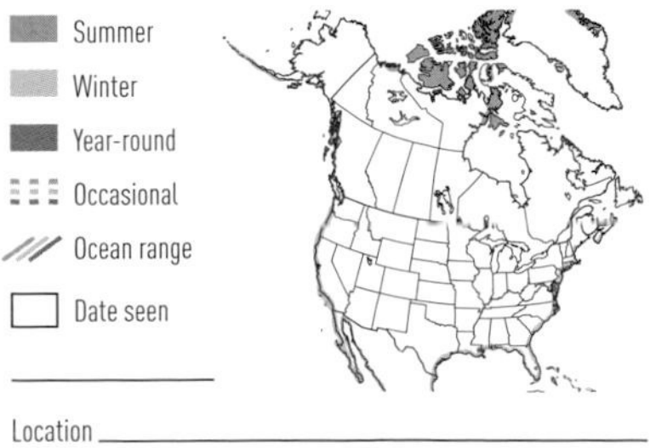

Calidris minutilla Length: 6"

LOOK FOR: A tiny (sparrow-sized!) shorebird, the Least Sandpiper can be hard to tell apart from the other small shorebirds known as peeps, which include the Western and Semipalmated Sandpipers. In good light, the Least's yellow legs and uniformly brown back are good field marks.

LISTEN FOR: A burry, rising *preep!* The Least's call is higher pitched than other peeps'. Peeps are named for their *peep* calls.

REMEMBER: Mixed flocks of peeps can be hard to sort out by species. Size differences are subtle. With practice you can sort out the Leasts from the other peeps by their slightly drooping fine-tipped bills and yellow legs.

WOW!

The Least Sandpiper is well named. It's the world's smallest sandpiper.

▶ *Least Sandpipers must watch out for the ever-hungry Laughing Gulls, which are large enough to swallow them whole.*

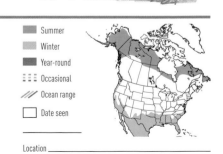

FIND IT: More common inland than other peeps, the Least Sandpiper can be found in small flocks on mud flats around lakes, ponds, and rivers. Leasts are methodical foragers that pick and probe for food in the mud.

Summer

Winter

Year-round

Occasional

Ocean range

Date seen ____________

Location ____________

SEMIPALMATED SANDPIPER

Calidris pusilla Length: 6¼"

LOOK FOR: The Semipalmated Sandpiper is another of our common (and hard to identify) peeps. Semis have straight, round-tipped bills and black legs. Winter adults are drab gray overall, while breeding adults and fresh-plumaged first-year birds have a browner overall appearance.

WOW!

Semipalmated Sandpipers get their name from their partially webbed toes. This is not a field mark, however, and it's a feature most birders never see.

LISTEN FOR: Flight call is a short *brip, brip,* lower and less whistly than the Least Sandpiper's. Also gives a loud *whe-do-do-do-do-do* call in flocks, which sounds like a mini car alarm.

▼ *Like other birds, Semipalmated Sandpipers can rest with half the brain asleep while the other half remains alert for danger.*

REMEMBER: The Semipalmated Sandpiper's bill is thicker than the Least Sandpiper's and has a stubbier point. In general, Semis look grayer overall than Leasts (which appear browner overall).

FIND IT: Found on mud flats and beaches, while traveling between its Arctic nesting grounds and South American wintering grounds, often in large flocks mixed with other peeps. Will forage in shallow water.

Summer

Winter

Year-round

Occasional

Ocean range

Date seen ______________

Location ______________

SPOTTED SANDPIPER

Actitis macularius Length: 7½"

LOOK FOR: This medium-sized shorebird retains its spots only during spring and summer. Fall and winter birds have dirty but unspotted breasts. The best field mark for the Spotted Sandpiper is its tail-bobbing walking style as it forages along the water's edge. Its white eye line is prominent and helps set the Spotted apart from the similar Solitary Sandpiper, which has a white eye-ring.

LISTEN FOR: Loud, ringing *peet-peet!* Often uttered in paired notes but in a longer series (*weet-weet-weet-weet*) when the bird is excited or startled.

REMEMBER: When spooked into flight, Spotties fly with a flap-flap-flap-sail rhythm, much like a Chimney Swift.

▶ *The Spotted Sandpiper's stiff-winged, stuttering flight is a good giveaway to its identity.*

WOW!

Another name for the Spotted Sandpiper is Teeter Peep, for the way it walks.

FIND IT: Scan along the edge of almost any pond, small stream, or muddy riverbank, and you are likely to see a Spotted Sandpiper. Spotties usually forage alone, sometimes in pairs.

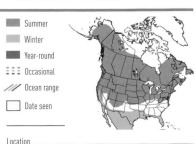

Summer
Winter
Year-round
Occasional
Ocean range
Date seen ______

Location ______

LOOK FOR: A football-sized bird, the American Woodcock is a shorebird that lives its life in and near woods. On the ground, it looks large bodied, big headed, with large black eyes and a long bill. Its eyes are set far back on the head, allowing it to watch for danger from behind and above as it probes for earthworms.

LISTEN FOR: The male's elaborate courtship display begins with a series of nasal *peent!* calls. Then he takes flight, wings whistling *too-too-too-too* as he flies upward in a spiral. As he tumbles back to earth, he twitters and burbles.

REMEMBER: Before you see a woodcock, you may hear the whistle of its wings as it flies past, or you may hear one *peent*ing at night.

▼ *An American Woodcock subdues a wriggling earthworm.*

WOW!

The woodcock's bill has a flexible, sensitive tip that can sense an earthworm and then open up just enough to grasp the worm and extract it from the ground.

FIND IT: Woodcocks favor wet woods for nesting and roosting, with nearby fields or meadows for nighttime foraging. The best time to find this species is in spring and summer, when males perform their "sky dance" display.

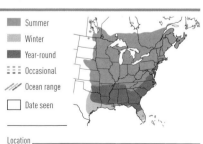

Summer

Winter

Year-round

Occasional

Ocean range

Date seen

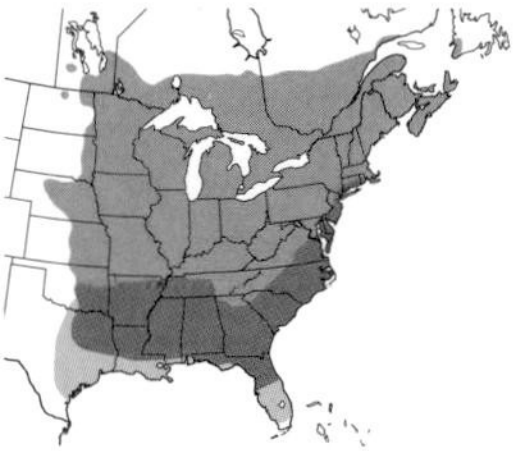

____

Location ______________

LOOK FOR: Dressed in the perfect camouflage for its grassy, marshy habitat, the Wilson's Snipe lurks, unseen by most birders. The stripes on the snipe's head and back help camouflage it.

LISTEN FOR: Our snipe has three sounds: Its startled-into-flight call is a harsh *skretch!* Its perched display song is a kestrel-like *ki-ki-ki-ki-ki.* And its courtship or winnowing display includes a long series of rising notes: *woo-woo-woo-woo*, a sound given in flight that is actually produced by the wind flowing over specialized outer tail feathers.

REMEMBER: Wilson's Snipe looks superficially like a dowitcher, but it is much less active and rarely found on open mud flats in flocks, as dowitchers are.

▼ *As a Wilson's Snipe performs its courtship display flight, it tilts from side to side, and the sound made by wind passing through its tail feathers changes tone.*

WOW!

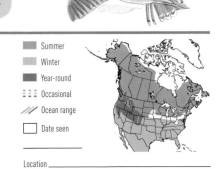

The snipe hunt is a practical joke named after the Wilson's Snipe. The victim is told to wait for a snipe to appear in a remote place and to use an empty sack and a stick to catch it.

FIND IT: Any muddy meadow or boggy wetland can host a snipe in migration. Before you see it, you may spook a snipe into a rapid zig-zagging flight. It may be gone before you recover your senses.

Summer
Winter
Year-round
Occasional
Ocean range
Date seen

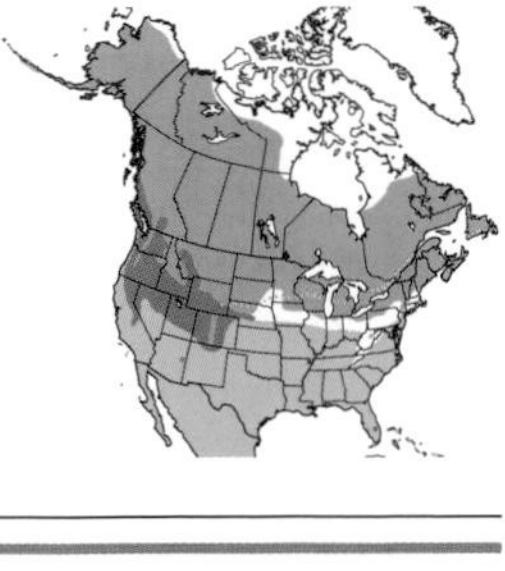

____

Location ________

LONG-BILLED DOWITCHER

Limnodromus scolopaceus Length: 11½"

LOOK FOR: Our two dowitcher species, Long-billed and Short-billed, are almost impossible to tell apart in the field. Both are chunky-looking birds, rich orange-brown in spring and summer, and gray overall in fall and winter. Both have long, straight bills. It's easier to tell them apart by where you find them (see Remember).

WOW!

It wasn't until the 1950s that ornithologists confirmed that Short-billed and Long-billed Dowitchers were two separate species.

LISTEN FOR: Calls are the best way to separate the two dowitchers. Long-billeds give a high-pitched *peep!* or *pee-deep!* Short-billeds say *tew-tew-tew*, usually in threes.

REMEMBER: *S* stands for Short-billed and salt water. Long-billeds are more commonly found in freshwater habitats (think *L* for Long-billed and lake).

► *Dowitchers like these Long-billeds feed by rapidly probing the mud. They look like feathered sewing machines.*

FIND IT: Look for dowitchers in large flocks on mud flats, shallow lakes, and ocean beaches (Short-billed), using their bills to probe for food. Long-billeds are more common inland during spring and fall migration.

Summer
Winter
Year-round
Occasional
Ocean range
Date seen

______

Location ______

Larus atricilla Length: 17"

LOOK FOR: A medium-sized and slender gull with a long dark bill and dark legs (both can be orange-red in breeding plumage). In spring and summer, the Laughing Gull's head is black. In winter, the head appears to be dirty white, but the gull retains the slate gray back and dark wingtips.

WOW!

Laughing Gulls will chase other birds to try to steal their food. They usually chase terns, which are smaller, but they will also harass birds as large as Brown Pelicans and Ospreys.

LISTEN FOR: This bird is named for its call, which sounds very much like someone laughing loudly, in either single notes—*Ahh! Ahhh!*—or in a series—*Ah-ah-ah-ah-ah!*

REMEMBER: Identifying gulls in nonbreeding plumage can be difficult. Overall size and shape is a good way to sort out nonbreeding-plumaged birds. Laughing Gulls appear slender overall and have a light, buoyant flying style.

► *A Laughing Gull harasses a Forster's Tern, trying to force it to drop its catch.*

FIND IT: Abundant along the Atlantic and Gulf coasts in the warm months, the Laughing Gull is a creature of saltwater habitats. Flocks can be seen foraging on beaches and loafing on dunes, piers, and breakwaters.

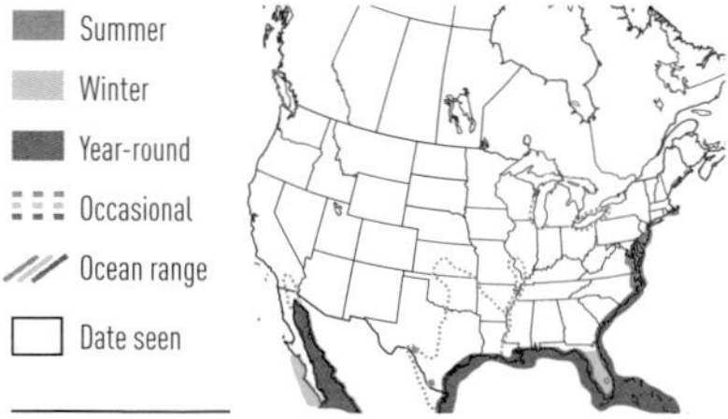

Date seen ____________

Location ____________

FRANKLIN'S GULL

Leucophaeus pipixcan Length: 14½"

LOOK FOR: The wingtips of Franklin's Gulls show a pattern of white-black-white. Breeding adult shows a black hood, a broken white eye-ring, and a red bill, and it may show rosy pink on the breast. Underwing is mostly white with black wingtips.

LISTEN FOR: Laughing cries similar to those of Laughing Gull, but higher pitched. A flock of calling Franklin's Gulls sounds like a bunch of teenage girls all crying *Ewwww!*

REMEMBER: This is the common black-hooded gull found on the northern Great Plains. Bonaparte's Gulls have white-tipped wings edged in black.

WOW!

Franklin's Gulls migrate way south and spend the winter along the Pacific Coast of South America!

▶ *Franklin's Gull flocks regularly follow tractors plowing fields. They scarf up insects disturbed by the plowing.*

FIND IT: Often found in spring and summer inland in large flocks in plowed or flooded fields, marshes, and lakes. Makes floating nest of vegetation on deep lakes. Migrates in flocks.

Summer
Winter
Year-round
Occasional
Ocean range
☐ Date seen

____

Location ______________

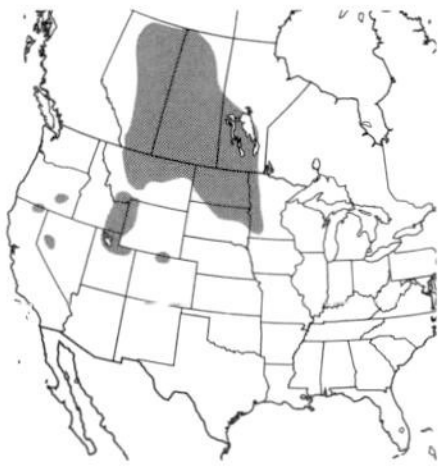

BONAPARTE'S GULL

Larus philadelphia Length: 13"

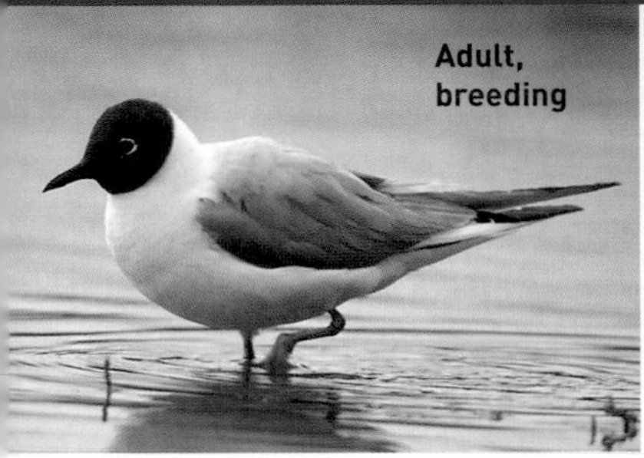

LOOK FOR: The Bonaparte's Gull is our smallest commonly encountered gull. It has a buoyant, ternlike flight, showing large white triangles on the leading edge of its gray wings. Black headed during spring and summer, in winter plumage the Bonie looks like it's wearing headphones.

LISTEN FOR: A very high-pitched, squally *nyahhh!* that sounds appropriate for a gull this small.

REMEMBER: The bouncy flight and flashes of white in the wings of the Bonaparte's Gull are field marks that work well, even in poor light or from a great distance.

WOW!

The Bonie is a picky eater. It does not visit garbage dumps for food as our other common gulls do.

▶ *"Pie slices" on the wingtips and "headphones" are good winter field marks for immature (left) and adult (right) Bonaparte's Gulls.*

FIND IT: Most commonly seen in winter along the coasts, but present inland during migration on rivers, lakes, and reservoirs. Often present in small flocks but rarely associating with other, larger gulls.

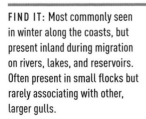

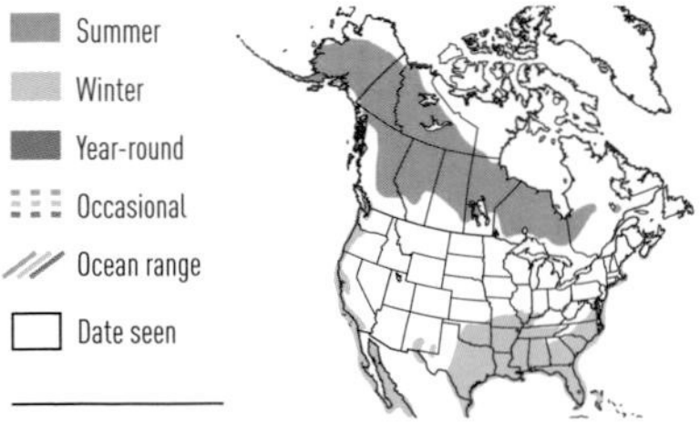

☐ Date seen ____________

Location ____________________________

WOW!

Heermann's Gulls are very aggressive birds. They often steal food from pelicans and other birds.

LOOK FOR: The breeding-plumaged adult's dark gray body, black legs and tail, bright white head, and deep red bill make this gull easy to identify. Nonbreeding Heermann's Gull has a gray head. In flight shows bold white and black tail.

LISTEN FOR: Nasal call is *awww!* Also high-pitched squeals.

REMEMBER: No other western gull has the combination of black legs and red bill.

▶ *Heermann's Gulls regularly pirate food from Brown Pelicans.*

FIND IT: Common within its coastal range, this species is dedicated to the ocean and rarely found far from the coast. Heermann's Gulls breed in early spring in Mexico and move north along the Pacific Coast in summer, following flocks of Brown Pelicans.

Summer

Winter

Year-round

Occasional

Ocean range

Date seen ________

Location ________

CALIFORNIA GULL

Larus californicus Length: 21"

LOOK FOR: Size-wise this gull is in between the larger Herring Gull and the smaller Ring-billed Gull and can be easily confused with either of them. Dark eyes and yellowish legs are key field marks for the adult California Gull. In flight it shows half-circles of white on black wingtips. Head is streaked with brown in nonbreeding plumage.

LISTEN FOR: Typical hoarse gull cries, but pitched higher than those of Herring Gull and deeper than those of Ring-billed Gull.

REMEMBER: Say it with me: "Medium-sized with dark eyes." Adults are easier than mottled first-year youngsters to separate from Herring and Ring-billed Gulls.

WOW!

If the Beach Boys had been birders, their hit song might have gone "I wish they all could be California Gulls!"

◄ *In 1848 California Gulls saved the crops of Mormon settlers in Utah when they devoured hordes of crop-eating grasshoppers.*

FIND IT: Nests on inland lakes and marshes and forages in farm fields, dumps, parks, and lakes, even in urban settings. Winters along the Pacific Coast, often resting on docks, on beaches, and in parks.

Summer
Winter
Year-round
Occasional
Ocean range
Date seen

____

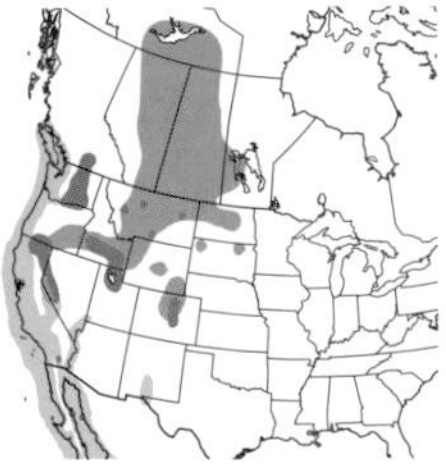

Location ______________

RING-BILLED GULL

Larus delawarensis Length: 17½–19"

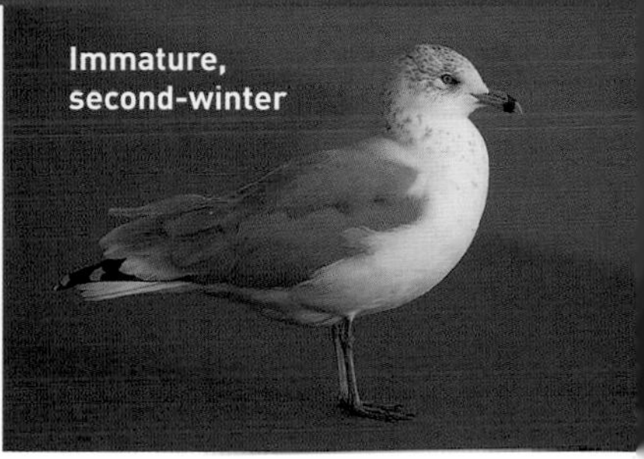

WOW!

The Ring-billed Gull's habit of hanging around fast-food restaurants has earned it the nicknames McGull, Dumpster Gull, and French-fry Gull.

LOOK FOR: Part of a group of large white-headed gulls, the Ring-billed Gull retains its namesake field mark (a clean, black ring on the bill) in all seasons, but that alone is not enough for positive identification because the Herring Gull also has a ring on the bill. Overall size, and bill size, is smaller than the Herring Gull, which is often confused with the Ring-billed.

LISTEN FOR: Basic call is a high-pitched, nasal, and squeally *klee-ear!*

REMEMBER: Ring-billed Gulls and Herring Gulls both have pale gray mantles (upper back and wings), white heads, and rings on the bills, but Ring-bills are nearly eight inches shorter and look shorter necked than Herrings.

▶ *A young Ring-billed Gull begs for a French fry.*

FIND IT: The Ring-bill is equally at home on ocean beaches, near freshwater reservoirs and dams, at landfills, near fast-food restaurants, and loafing in a large parking lot. More common inland than other gull species.

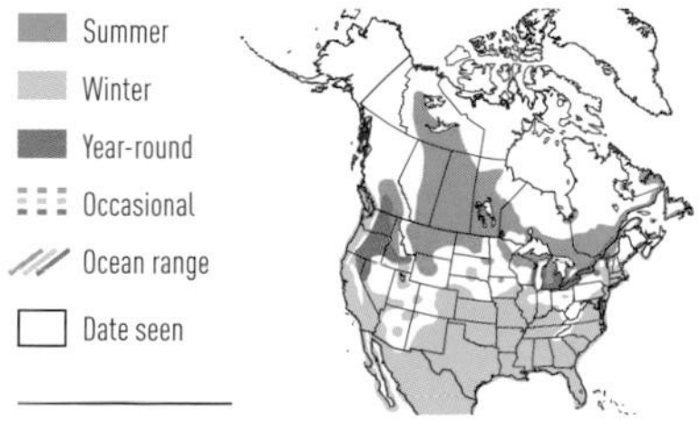

Date seen ______

Location ______

HERRING GULL

Larus argentatus Length: 25"

LOOK FOR: Our most common large gull continent-wide, the Herring Gull is heavy bodied and heavy billed. In summer adult plumage, the head is clean white, the bill is yellow with a red spot, and the legs are pinkish. Winter adults have "dirty" heads.

LISTEN FOR: Call is similar to other common gulls but a lower-pitched and hoarser two-syllabled *oww-uh! oww-uh!* Also utters a rapid, tuneless *uh-uh-uh.*

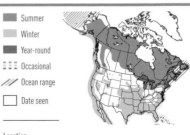

▲ *A Herring Gull prepares to drop a clam on the hard asphalt. When the shell breaks, the gull will descend to eat.*

REMEMBER: Herring Gulls take four years to reach adult plumage. They start out all brown and get whiter and cleaner-looking as they age.

WOW!

Herring Gulls sometimes carry large shellfish such as mussels, clams, and oysters in the air and drop them onto a hard surface in order to break them open and eat them.

FIND IT: Herring Gulls are usually found near large bodies of water, though they also visit large landfills. In winter, they can be found on inland rivers, particularly near dams.

Summer
Winter
Year-round
Occasional
Ocean range
Date seen ______

Location ______

Larus glaucescens Length: 26"

Adult (center), with juveniles

LOOK FOR: Named for its pale appearance, this very large gull has pinkish legs, a bold yellow bill, and medium gray back and wings edged in white.

LISTEN FOR: This species has two very different calls. One is typical gull-like screams with a few nasal *ka-ka-kas* thrown in. The other is a piercing cry that sounds like a bull elk trying to imitate a Common Loon.

WOW!

The Glaucous-winged Gull is not choosy about its mates. It hybridizes with several other large gulls where their ranges overlap.

REMEMBER: Compared with our other large pale gulls, the Glaucous-winged looks fairly plain and unpatterned overall. Its wingtips match its back in color.

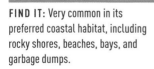

▶ *Glaucous-winged Gulls often nest on rooftops in coastal towns along the Pacific.*

FIND IT: Very common in its preferred coastal habitat, including rocky shores, beaches, bays, and garbage dumps.

Summer

Winter

Year-round

Occasional

Ocean range

Date seen ________

Location ________

WESTERN GULL

Larus occidentalis Length: 25"

LOOK FOR: The Western Gull is the only common large dark-backed gull in its range. The heavy yellow bill stands out. Adult has a dark gray back and pink legs. First-year birds are all dark brown, becoming more cleanly marked as they reach breeding age in the fourth year.

LISTEN FOR: Low, hoarse calls that sound like an angry Chihuahua's barks.

REMEMBER: Western Gull is a big bird with a dark back and a huge yellow bill. Heermann's Gull is black backed but much smaller.

WOW!
These birds boldly hang around California sea lion colonies, waiting to scavenge dead pups and other yummy things.

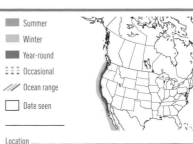

▲ *Western Gulls often nest near other breeding birds so they can steal their eggs and food.*

FIND IT: Common along the Pacific Coast at all seasons, but rare inland. Found on beaches, docks, breakwaters, parks, parking lots, and dumps.

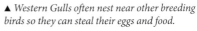

Summer
Winter
Year-round
Occasional
Ocean range
Date seen ____________

Location ______________________________

GREAT BLACK-BACKED GULL

Larus marinus Length: 30"

WOW!
The Great Black-backed Gull can be confused with a Bald Eagle when seen from a distance. But the gull has a white head and breast, while an adult Bald Eagle has just a white head.

LOOK FOR: The Great Black-backed Gull is a huge bird, and its size alone is often enough to pick this species out of a flock of loafing gulls. Big headed with a massive bill, the Great Black-backed dwarfs other common gulls. The dark back of adult birds is an excellent field mark, as are the pink legs. Young birds have checkered backs and wings for their first two years.

LISTEN FOR: Common call is a deep, nasal *owwh! owwh!* But also utters a variety of honks and chortles.

REMEMBER: This is the only gull in the East with the combination of black back and pink legs. In the West, there are other gulls with these features.

▼ *Great Black-backed Gulls often act like hawks, preying on small seabirds such as puffins.*

FIND IT: Very common along the entire East Coast, usually near salt water. In winter Great Black-backs can be found inland on the Great Lakes and near landfills and reservoirs.

Location ______

FORSTER'S TERN

Sterna forsteri Length: 13–14½"

Adult, breeding

Adult, nonbreeding

LOOK FOR: If you could see a Forster's Tern perched next to a Common Tern, you'd see that the Forster's is slightly larger and bulkier with a longer, orange (not red) bill. In spring and summer, the Forster's underparts are white (not gray), and the tail usually extends beyond the tips of the folded wings on resting birds.

LISTEN FOR: A harsh, nasal *kyerrr! kyerrr!* given both singly and in a series. Calls are buzzier than those of the Common Tern.

REMEMBER: Winter adult and juvenile Forster's Terns lack the dark shoulder bars of the Common Tern. They also have black raccoon-mask eye patches, but this black does not wrap around the back of the head.

WOW!

Early ornithologists didn't recognize Forster's and Common Terns as two separate species. (They didn't have the benefit of our modern optics and field guides, so we should cut them some slack.)

◀ *In winter, the Forster's Tern (left) sports a bandit mask and white wings, while the Common Tern (right) has dark wingtips, nape, and shoulders.*

FIND IT: Forster's Terns love marshes and can be found in both freshwater and saltwater habitats, including bays, lakes, reservoirs, and rivers. Many Forster's Terns winter along the southern coasts of the U.S.

Summer
Winter
Year-round
Occasional
Ocean range
Date seen ____________

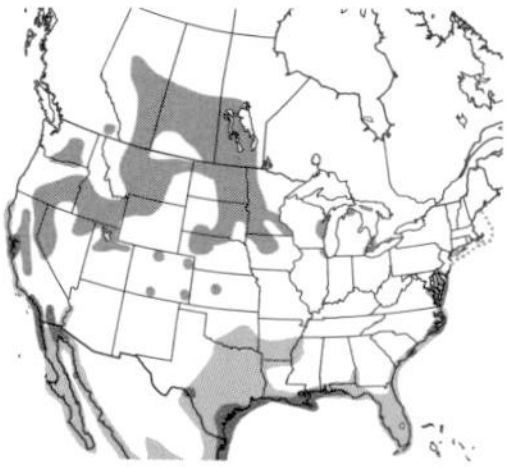

Location ____________

LOOK FOR: The Common Tern is a medium-sized tern with a look-alike cousin in the Forster's Tern. In breeding plumage, the Common has a thicker, redder bill and a gray body (the belly of the Forster's is all white). In winter and juvenal plumages, the Common Tern has a dark shoulder bar and a wraparound black patch behind the eyes.

LISTEN FOR: A loud, burry *kee-yarr, kee-yarr!*

REMEMBER: In winter, adult and juvenile Common and Forster's Terns are very similar, but the Forster's Terns lack shoulder bars and their black eye patches do not connect on the back of the head.

▶ *Common Terns are specialists at fishing in shallow coastal waters, spotting a fish, then plunging, bill-first, to capture it.*

WOW!

If you get too close to a Common Tern nesting colony, expect to get dive-bombed by angry terns. They might even decorate you with a splat of poop if they really want you to leave.

FIND IT: In the Northeast, the Common Tern is the most common tern, but elsewhere the Forster's is often more common. Preferred habitat of Common Terns is ocean bays, beaches, lakes, and large rivers.

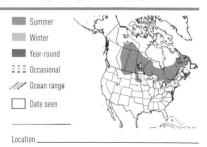

Summer

Winter

Year-round

Occasional

Ocean range

☐ Date seen

____

Location __________

ELEGANT TERN

Thalasseus elegans Length: 17"

LOOK FOR: A medium-sized tern, white bodied with a shaggy black crest and a thin orange-red bill. Black crown covers the forehead on breeding adults. Nonbreeding-plumaged birds show a white forehead.

LISTEN FOR: Loud single call is *keek!* Two-note call is a burry *karr-rick!*

REMEMBER: Compared with the similar Royal Tern, the Elegant Tern is smaller, with a thinner, more delicate bill. You could even say it's more elegant-looking!

WOW!

Elegant Terns love anchovies. When the anchovy population rises, this species enjoys greater nesting success.

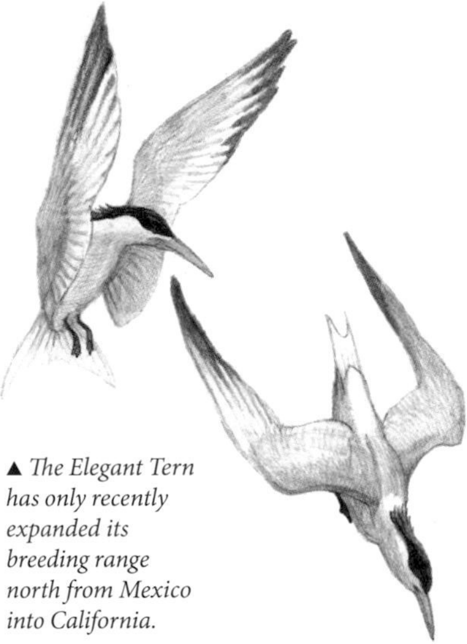

▲ *The Elegant Tern has only recently expanded its breeding range north from Mexico into California.*

FIND IT: A coastal bird, the Elegant Tern forages in shallow waters in bays and estuaries. After breeding, these birds move north along the coast for the summer.

Date seen ______

Location ______

CASPIAN TERN

Sterna caspia Length: 21"

Adult, breeding

LOOK FOR: The Caspian Tern is a large, gull-sized tern with a heavy red-orange bill. In summer, its striking crown is so dark, it is almost impossible to see its eye. Winter and juvenile Caspians have dirty foreheads.

LISTEN FOR: Utters a raspy, ducklike *ahrakk!* Begging call of juvenile is a high-pitched, wavery whistle.

REMEMBER: The Caspian Tern's look-alike relative is the Royal Tern, a bird that is rarely found inland. One field guide calls the Royal's bill "carrot-orange." So if you can remember that the Royals live at the beach (like royalty) counting the "carrots" in their diamonds, you can sort them out from the much less royal Caspians.

WOW!

Young Caspian Terns beg food from their parents months after they are able to fend for themselves. But don't try this with your parents. Trust me, it doesn't work.

◄ *A young Caspian Tern begs noisily from its parent.*

FIND IT: Our most common large tern found inland on lakes and along large rivers, but also along both coasts. Look for them among flocks of gulls on beaches. Their bright bills will stand out in the crowd.

Summer

Winter

Year-round

Occasional

Ocean range

☐ Date seen

____

Location ______________

COMMON MURRE

Uria aalge Length: 17–17½"

LOOK FOR: A duck-sized, football-shaped bird that is dark above and white below, with a slender, pointed black bill. Flies very fast, with rapid changes of direction.

LISTEN FOR: This species is named, apparently, for the low, moaning sound (*mrrrrrr*) it makes at the nest burrow.

REMEMBER: Thick-billed Murres are very similar, but their range is far more northerly. Their bills are also thicker and shorter.

▼ *Murres are very fast fliers but even better divers and swimmers.*

WOW!

Common Murres, like other alcids, are closely related to penguins, except that, unlike penguins, murres can fly.

FIND IT: Most common near nesting colonies on rocky ocean cliffs and offshore islands. Often seen flying to and from colonies, bringing food to young. Winters at sea.

Summer

Winter

Year-round

Occasional

Ocean range

Date seen ______

Location ______

Rynchops niger Length: 18"

LOOK FOR: One of our most distinctive birds, the Black Skimmer is named for its method of foraging—its elongated lower bill (mandible) plows lightly through the water, feeling for small fish. The skimmer's very long body and bill make its shape easy to pick out in a flock of resting gulls and terns.

LISTEN FOR: Skimmers utter a high-pitched, burry bark: *eerrff! eerrff!*

REMEMBER: Skimmers are very graceful fliers, wheeling in tight turns and gliding just above the water's surface. Standing at rest, they appear to lack eyes—their black eyes and black hoods blend together completely.

WOW!

The Black Skimmer's bill is thin and knifelike, perfect for slicing through the water, skimming for tasty fish. The skimmer does not usually see its food. It feels for its food as it skims along.

▶ *When the lower mandible contacts a fish, the skimmer snaps its bill shut.*

FIND IT: Skimmers prefer to feed on and nest near smooth salt water, such as estuaries, bays, and lagoons, where the calm water makes foraging easier. Flocks resting on beaches will all face into the wind.

Summer
Winter
Year-round
Occasional
Ocean range
Date seen

Location ______

ATLANTIC PUFFIN

Fratercula arctica Length: 12½"

LOOK FOR: Even people who have never seen a puffin know what a puffin looks like, with its huge rainbow-colored bill, white face, orange feet, and handsome tuxedo plumage. Winter adults have very dull bills and gray faces.

LISTEN FOR: At their breeding colonies, Atlantic Puffins give calls that sound *exactly* like chain saws. It's worth the trip to their nesting islands just to hear this.

REMEMBER: The Atlantic Puffin is the only puffin found on the East Coast. Two other puffin species, the Horned Puffin and the Tufted Puffin, are found along the West Coast.

WOW!

The Atlantic Puffin nearly became extinct because of overhunting of the birds and harvesting of eggs. The population has increased dramatically in the past 30 years through conservation efforts.

▶ *Conservationists used decoys and sound recordings to attract Atlantic Puffins back to rocky islands where they once nested.*

FIND IT: Atlantic Puffins are rarely seen from shore. The best way to see them is to take a charter-boat trip to one of their protected nesting islands off the coast of northern New England.

Summer
Winter
Year-round
Occasional
Ocean range
Date seen

Location ______________________

TUFTED PUFFIN

Fratercula cirrhata Length: 15–16"

LOOK FOR: A dark-bodied seabird with a massive orange bill. Breeding-plumaged adult has a white face and bright cream-colored tufts on the back of the head.

LISTEN FOR: Growls in low tones near nest site, but otherwise silent.

REMEMBER: Adults in summer are unmistakable. Young birds and winter adults are gray bodied with greatly reduced orange on the bill.

WOW!

Tufted Puffins look very colorful in summer and very dull in winter.

◀ *Tufted Puffins dig their nesting burrows in soil on cliffs near the ocean. If no soil is present, they use a rocky crevice.*

FIND IT: Uncommon in open ocean. Most sightings are at nesting colonies on rocky ocean cliffs and steep grassy slopes on offshore islands.

Summer

Winter

Year-round

Occasional

Ocean range

Date seen ______

Location ______

LOOK FOR: Named for its pigeonlike appearance, this small alcid is black in the breeding season with large white wing patches, a pointed black bill, and bright orange legs. In nonbreeding plumage the bird is pale gray below with black wings.

LISTEN FOR: Gives shrill peeps and whistles near the breeding colony.

REMEMBER: Look for the large white wing patches, each bisected by a black bar.

WOW!

Pigeon Guillemots catch their food (fish, shrimp, crabs) by diving underwater and swimming in pursuit, using both their wings and feet.

▶ *Most Pigeon Guillemots nest in colonies on rocky cliffs that are inaccessible to predators.*

FIND IT: Because they nest colonially on sea cliffs, Pigeon Guillemots can be seen flying to and from their nests during the spring and summer months. They prefer nearshore waters, unlike most members of the Auk family.

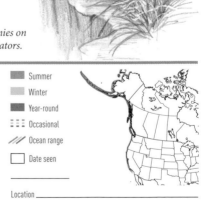

Summer
Winter
Year-round
Occasional
Ocean range
Date seen

____

Location ______________

LOOK FOR: Larger and longer than a Rock Pigeon with a white half-collar on nape of neck and a pale gray band on the end of the tail (for which it is named).

LISTEN FOR: Call is a very owl-like *whooo-hoo.*

REMEMBER: The Band-tailed Pigeon often lands in trees to forage. Rock Pigeons prefer to land on buildings and other structures.

WOW!
When a Band-tailed Pigeon takes off suddenly, its wings make a loud clapping sound.

◀ *Band-tailed Pigeons are adept at clambering around in trees to get at food.*

FIND IT: Prefers woods containing oaks in mountains, foothills, and canyons, where it forages for fruits, nuts, and berries. Often found in flocks.

Summer
Winter
Year-round
Occasional
Ocean range
☐ Date seen

____

Location ______________

EURASIAN COLLARED-DOVE

Streptopelia decaocto Length: 13"

LOOK FOR: Looking like a washed-out Mourning Dove wearing a black bandanna around its neck, the Eurasian Collared-Dove is a recent immigrant to North America. Chunky and pale overall, it lacks the MoDo's long tapered tail.

LISTEN FOR: Eurasian Collared-Doves repeat *coo-COO-coo* over and over. The call sounds like a Mourning Dove with a sore throat stuck on Repeat.

REMEMBER: In flight, the Eurasian Collared-Dove looks chunkier than a Mourning Dove and lacks the MoDo's long central tail feathers.

▼ *Courtship, mating, and many other aspects of the adaptable Eurasian Collared-Dove's life are carried out on power lines.*

WOW!

From an accidental release in the Bahamas in 1974, the Eurasian Collared-Dove has invaded the southeastern U.S. It is so adaptable that it may become one of our most common birds.

FIND IT: Introduced from Europe, the Eurasian Collared-Dove is colonizing North America at an alarming rate. Common in suburban settings, where it likes to perch on power lines and other exposed places.

Summer

Winter

Year-round

Occasional

Ocean range

Date seen

____

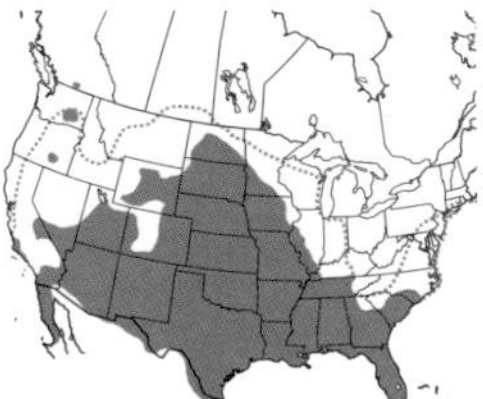

Location ______________

LOOK FOR: The bird formerly known as Rock Dove and Feral Pigeon—and still known by lots of other, less pleasant names—comes in a variety of plumage colors, but the most common is gray bodied and dark headed with orange feet and orange eyes.

LISTEN FOR: Rock Pigeons make a low, gurgling *urr-cooooo!*, most often at roosts or nest sites. When Rock Pigeons explode into flight, their wings make a loud slapping sound.

WOW!

Rock Pigeons have helped in the reestablishment of the Peregrine Falcon. Peregrines that nest in cities eat lots of plump and tasty Rock Pigeons.

REMEMBER: Rock Pigeons are strong and direct fliers, and they can be confused with falcons in flight. A closer look shows the pigeon's potbellied, broad-winged, small-headed appearance to be different from a falcon's flight shape.

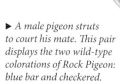

▶ *A male pigeon struts to court his mate. This pair displays the two wild-type colorations of Rock Pigeon: blue bar and checkered.*

FIND IT: Widespread and common, especially in cities, on bridges, in parks, and near farms with open barns, the Rock Pigeon was introduced to North America from Europe, where its native relatives nest on rocky cliffs.

Summer

Winter

Year-round

Occasional

Ocean range

☐ Date seen

____

Location ________________________

LOOK FOR: Named for large white wing patches, which are most obvious in flight, as they contrast with dark flight feathers. Perched birds show a long white wing edge. The tail is rounded with white outer tips.

LISTEN FOR: Call sounds very similar to Barred Owl's *who cooks for YOU!* When you hear a day-calling "owl," it may be this species.

REMEMBER: Similar Mourning Dove has a pointed (not rounded) tail and no white in the wings.

WOW!
The White-winged Dove drinks nectar and rainwater from cactus flowers and in doing so helps to spread pollen among cacti.

▲ *Most White-winged Doves head south in winter.*

FIND IT: A dove of the Southwest that prefers woods along rivers, groves, brushland, and other semi-open habitats. Increasingly found in towns and parks. Range is expanding northward.

Summer
Winter
Year-round
Occasional
Ocean range
Date seen ______

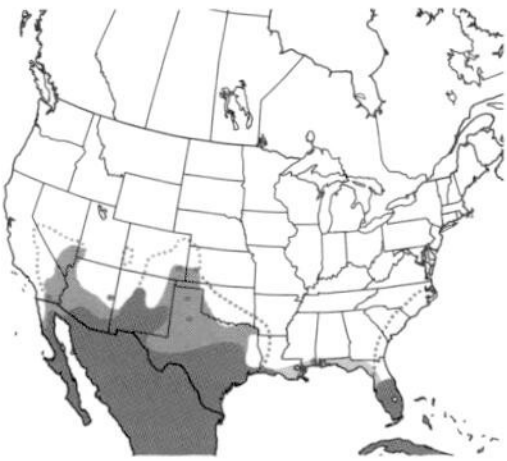

Location ______

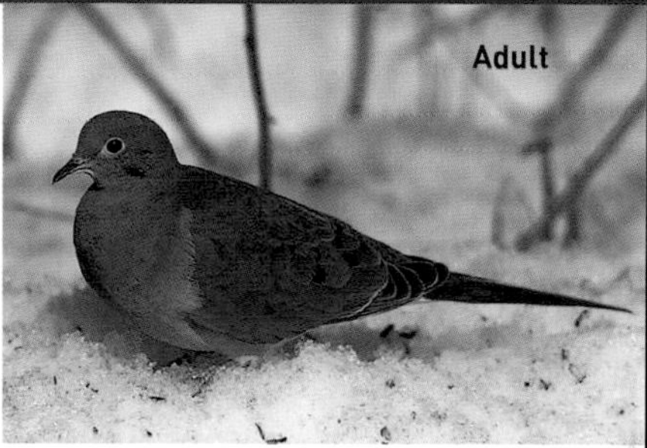

LOOK FOR: The slender brown shape of the Mourning Dove, with its long tapered tail, is a familiar sight all across North America. When perched, the dove shows black spots on tan wings, and the adult can show glistening shades of purple, pink, and green on the neck and head. In flight, the Mourning Dove's tail feathers show white tips.

LISTEN FOR: This species is named for its sad-sounding cooing: *ah-ooh! whoo-whoo-whoo*. Nonbirders often confuse the Mourning Dove's call with an owl's hooting.

REMEMBER: Their rapid flight makes the MoDo look a little like a hawk, but the wedge-shaped dove tail with white spots plus the dove's small head should clinch the ID.

▼ *Mourning Doves build the flimsiest of nests, often in weird places.*

WOW!

The Mourning Dove has a built-in straw! Other birds have to scoop water in their bills and tilt their heads back to swallow. But the MoDo can drink water by sucking it up through its bill.

FIND IT: You can find Mourning Doves almost anywhere except in thickly wooded habitats. They come to bird feeders and often perch on wires and forage on lawns and gravel roads; they also form huge flocks in agricultural fields during fall and winter.

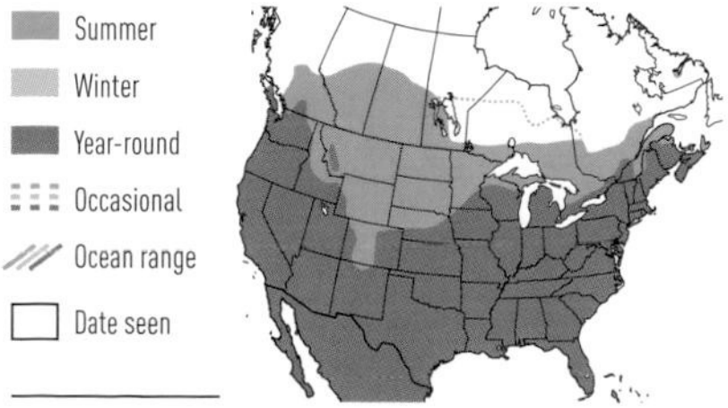

Location ________

YELLOW-BILLED and BLACK-BILLED CUCKOOS

Coccyzus americanus, Coccyzus erythropthalmus Length: 12"

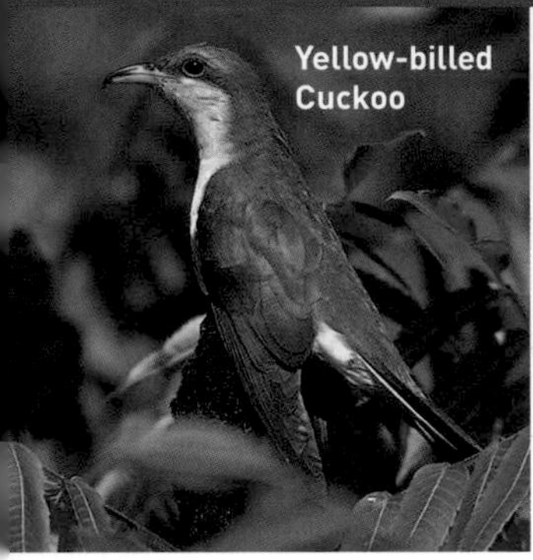

LOOK FOR: This long and slender skulker can be hard to see despite its size. In flight, the Yellow-billed shows large rusty patches in its wings and appears very long tailed. The similar Black-billed Cuckoo has a smaller black bill and a red eye-ring, and is cream colored below and olive above, with no rust color in the wings.

WOW!

Because the Yellow-billed Cuckoo commonly sings on humid summer afternoons, the bird's folk name is Rain Crow, as it is said to predict a coming rainstorm.

LISTEN FOR: From deep cover, Yellow-billed Cuckoos give a hollow, hoarse, two-syllabled *tee-oo, tee-oo, tee-oo* in a series and a mechanical-sounding *kik-kik-kik-kik-kaKOW, kaKOW.*

REMEMBER: A cuckoo's flight shape is long and slender. If you see lots of white below, white spots in the tail, and rust in the wings, you've got a Yellow-billed. If the bird is plainer, it's a Black-billed.

▶ *Yellow-billed Cuckoos have specialized stomachs that allow them to eat hairy caterpillars no other birds will touch.*

FIND IT: Cuckoos are more often heard than seen. The Yellow-billed is a bit less secretive than the Black-billed, which prefers woodlands near water. Both species like thick foliage where they can find caterpillars.

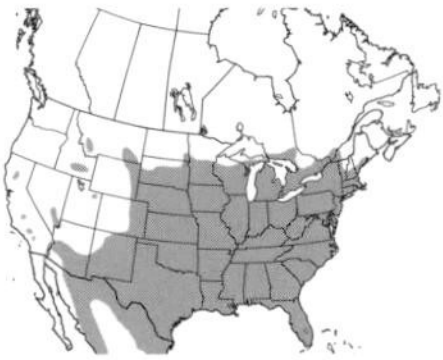

YELLOW-BILLED CUCKOO

☐ Date seen ______________

Location ______________

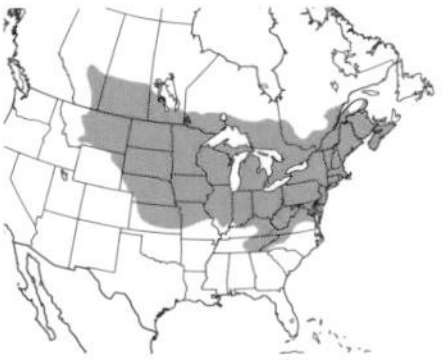

BLACK-BILLED CUCKOO

☐ Date seen ______________

____

Geococcyx californianus Length: 23"

Adult

LOOK FOR: This large streaky bird with a shaggy head is unmistakable, though it does not look much like the famous cartoon. The Greater Roadrunner has a long tail spotted with white and a head with a crest that it raises and lowers as it stalks through the dry open country while hunting.

LISTEN FOR: A very dovelike series of coos, *cu-cu-cu-cu-cuurrrr*, that slows down and drops in tone as it ends. No Greater Roadrunner has ever been recorded as saying *meep-meep!*

REMEMBER: Roadrunners are fast birds when chasing prey—they can run as fast as 15 mph. They prefer to walk or run on the ground rather than flying and will fly only as a last resort.

WOW!

Roadrunners eat almost anything they can catch: lizards and snakes, small rodents, scorpions and tarantulas, and large insects. They'll even leap up to catch hummingbirds at nectar feeders.

▶ *Far from its cartoon image as prey to the coyote, the Greater Roadrunner is a wily and voracious predator itself.*

FIND IT: Greater Roadrunners prefer very dry habitats, such as brushy desert and chaparral. Watch for them moving methodically through a habitat, hoping to startle a lizard, snake, or mouse into revealing itself.

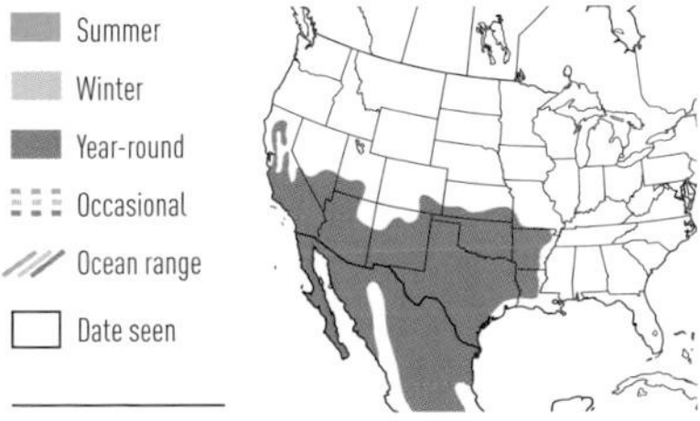

____

Location ______________

LOOK FOR: This bird could be called the No-eared Owl, since its ear tufts are almost impossible to see. Its warm brown color, round face, and flopping, moth-like flight are better field marks. Because it often hunts during daylight hours in open country, the Short-eared Owl is easier to see than our other owls.

LISTEN FOR: Gives a hoarse, scraping bark in flight. Also utters a series of short *kek-kek-kek-kek* sounds.

REMEMBER: In flight, the Short-eared Owl might be confused with the Long-eared Owl or Barn Owl. Barn Owls are white below and much paler overall. Short-ears show large buffy patches on their upper wings.

▼ *Hunting by sight and sound, a Short-eared Owl flies low over a meadow on a winter afternoon.*

WOW!

Short-eared Owls can clap their wings! Males use the wing clap as part of their courtship display, and both adults use it to scare predators and intruders away from the nest.

FIND IT: Short-eared Owls are grassland-loving birds, and where the hunting is good, multiple birds may be found, especially in winter. They use tundra, prairies, old-fields, and recovering habitats such as strip mines.

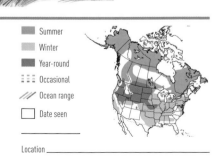

Summer

Winter

Year-round

Occasional

Ocean range

☐ Date seen

____

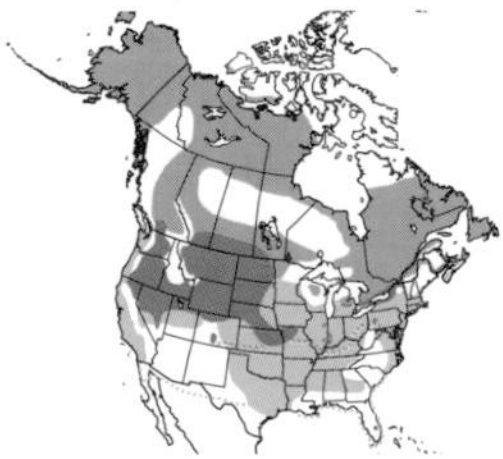

Location ______________

LOOK FOR: The heart-shaped white face, brilliant white underparts, and cinnamon brown back make the Barn Owl one of our most distinctive night birds.

LISTEN FOR: A shrieking, downward hiss that sounds like a woman screaming in terror. You are more likely to hear a Barn Owl's call on a dark night than to see one, and once you do you'll never forget it.

REMEMBER: The Barn Owl's low, mothlike flight might be confused with the flying styles of the Short-eared Owl or the Northern Harrier, but only the Barn Owl is all white underneath—giving it a ghostly appearance the others lack.

▼ *Barn Owls are great friends to the farmer. They will clear a barn of rats and mice.*

WOW!

The feathers on the face of the Barn Owl form a heart-shaped bowl. This shape channels sound waves to the owl's ears, helping it to locate prey even on the darkest nights.

FIND IT: Barn Owls are in decline over most of their range because of loss of habitat and adequate nesting sites (old barns, hollow trees). Look for them at dusk and dawn, flying low to the ground as they hunt in open areas.

Summer
Winter
Year-round
Occasional
Ocean range
Date seen

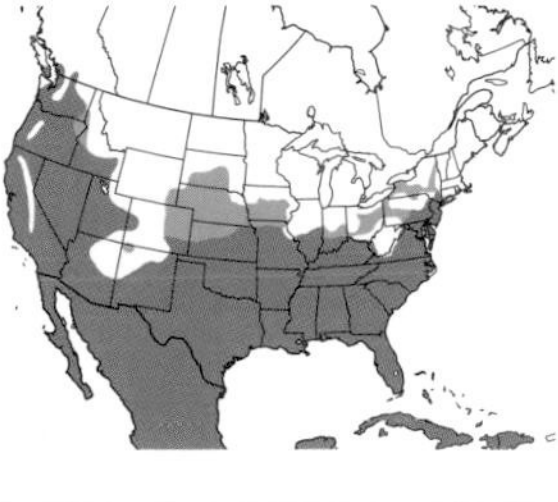

Location

GREAT HORNED OWL

Bubo virginianus Length: 23"

LOOK FOR: This huge nighttime predator is a common, year-round resident all across North America, nesting in woodlands and hunting woodland edges. Its large size, tufted round head, and horizontally striped breast and belly are excellent field marks.

LISTEN FOR: Deep, booming pattern of five hoots, usually *whoo-who-who, whooo-whoo.* Great Horneds begin courtship calling in midwinter, so this is a good time to listen for their far-carrying calls.

REMEMBER: Few other raptors look as large and bulky as the Great Horned Owl. Their big-headed, horned shape is distinctive.

WOW!

Although the female is larger than the male, he has a deeper voice and more complex call. If you hear two Great Horneds calling to each other, the voice with the deeper, lower notes belongs to the male.

◄ *Great Horned Owls are about the only predators that kill and eat skunks.*

FIND IT: During the day, Great Horneds roost in deep cover but may be harassed loudly by groups of crows. They are best located by sound, but at dawn and dusk they often perch up high in trees, on towers, and on buildings.

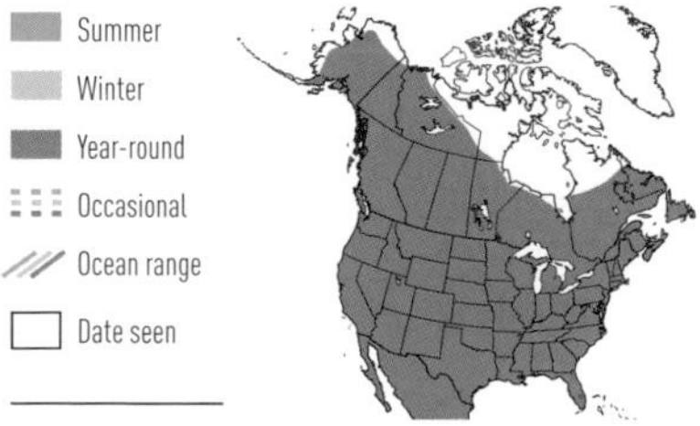

☐ Date seen

____

Location ______________________________

SPOTTED OWL

Strix occidentalis Length: 17½–18"

LOOK FOR: The Spotted Owl is a medium-sized, round, dark brown owl with dark eyes. It gets its name from the large white spots on its belly.

LISTEN FOR: Loud, high-pitched hoots in groups: *whoo-whowho-WHOO.*

REMEMBER: The Spotted Owl has spots on the breast. The very similar Barred Owl has bold dark vertical streaks on its breast.

◀ *A Spotted Owl in its daytime roost in a tree.*

WOW!

Spotted Owls hunt mostly at night, capturing small mammals on the ground and even catching bats in the air.

FIND IT: Rare in old-growth forests of the mountain West and pine-oak forested canyons of the Southwest. Most often seen while roosting during daylight. Most birds are resident but some at high elevations move lower in winter.

Summer

Winter

Year-round

Occasional

Ocean range

Date seen

Location

LOOK FOR: Mostly nocturnal and heard far more often than it is seen, the Barred Owl may be active on cloudy days, particularly when feeding hungry nestlings in late winter and spring.

LISTEN FOR: *Who cooks for you! Who cooks for you-all!* The Barred Owl's low-pitched hooting call is one of the better known among birders. Barred Owls will call at any time of year, and they will even call during the day in response to a loud noise.

REMEMBER: Among the three common owls of eastern woodlands, the screech-owl has the highest voice and the Great Horned the lowest. The Barred Owl's call is pitched between these two and always follows its distinctive eight-note pattern.

WOW!

In western North America, the Barred Owl is expanding its range, and in some areas this is forcing out the closely related but much smaller and rarer Spotted Owl.

▶ *Barred Owls will wade in creeks to catch crayfish to eat.*

FIND IT: Most common in the Southeast but present across eastern North America in mixed woodlands, especially woods near water.

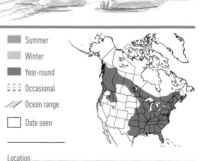

Summer

Winter

Year-round

Occasional

Ocean range

Date seen ________

Location ________

Strix nebulosa Length: 26–28"

LOOK FOR: A huge dark owl with a round face and piercing yellow eyes. Streaking on breast is smudgy. Black-and-white pattern on chin is distinctive. Pale feathers between eyes and bill form a large gray X.

LISTEN FOR: A series of ten or fewer deep-toned hoots, which get softer near the end.

REMEMBER: The Great Gray has a large, rounded head, unlike the more common Great Horned Owl's, which shows ear tufts.

WOW!

Great Gray Owls appear huge, but if you stripped all their feathers off, you'd see that their bodies are actually much smaller than you'd think.

▸ *Using their facial disks to focus sounds to their ears, Great Gray Owls can home in on their prey by sound. They may try to capture prey they hear deep below the surface of the snow.*

FIND IT: Uncommon in dense boreal forest and nearby bogs and meadows in the North. Found in mountain clearings and burned-over woods in the West. Active hunter in both day and night.

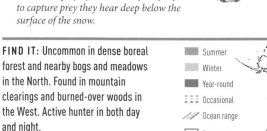

Summer

Winter

Year-round

Occasional

Ocean range

Date seen ____________

Location ____________

SNOWY OWL

Bubo scandiacus Length: 24"

LOOK FOR: This nearly all-white bird of the far North is difficult to confuse with anything else. Young birds and adult females are streaked with black. Older adult males are pure white. Some winters when small mammals are scarce in the far North, Snowy Owls head south in search of food.

WOW!

In the Arctic in summer, the sun never goes down, so Snowies are more adept than other owls at daytime hunting. They may catch and consume as many as 1,600 lemmings in a single year.

LISTEN FOR: Snowies are usually silent while wintering south of the Arctic nesting grounds, but if you spend the summer on the Arctic tundra, you'll probably hear a Snowy Owl make some noise.

REMEMBER: Snowy Owls that come south in winter may be stressed and hungry, so it's best to enjoy watching them from a distance, where you won't disturb them.

▶ *A male Snowy Owl offers his mate a prey item, hoping to impress her.*

FIND IT: Every winter some Snowy Owls journey south from the Arctic tundra in search of food, and birders see them perched on fenceposts along meadows and coastal dunes and at other large, open expanses.

Summer
Winter
Year-round
Occasional
Ocean range
Date seen

____

Location ___________

EASTERN SCREECH-OWL

Megascops asio Length: 8½"

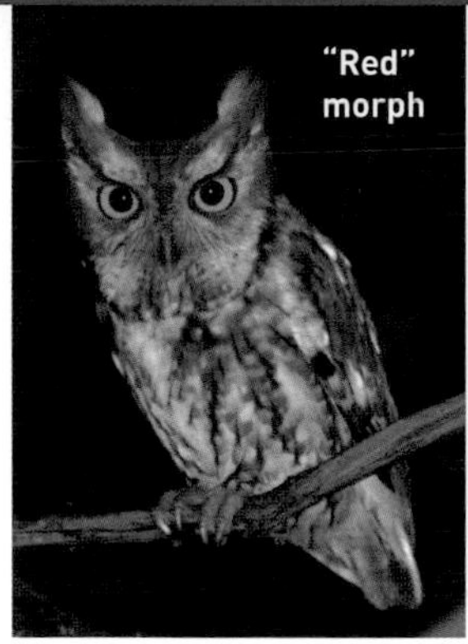

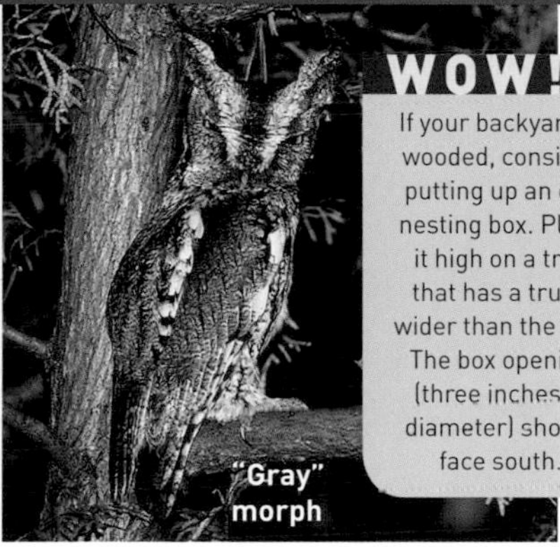

WOW!

If your backyard is wooded, consider putting up an owl nesting box. Place it high on a tree that has a trunk wider than the box. The box opening (three inches in diameter) should face south.

LOOK FOR: The most common of our small owls, the Eastern Screech-Owl comes in two color versions, or morphs: red and gray, with gray being the more common. Its small size, cryptic coloration, and inactivity during the day make this bird easy to overlook.

LISTEN FOR: Eastern Screech-Owls utter a series of high, wavery whinnies that descend in tone. This is often followed by a long trill on a single, lower tone.

REMEMBER: Your first clue to a screech-owl's presence might be hearing its haunting call at dusk. If it sounds close by, try slowly moving a flashlight across nearby trees, watching for the reflection of the bird's yellow eyes.

▶ *An Eastern Screech-Owl peeks out of a nest box to soak up the warmth of the afternoon sun.*

FIND IT: This species can be difficult to find because it is nocturnal. A cavity nester, it prefers habitats with old trees that have holes to nest in. Found in woodland settings, including suburbs and city parks.

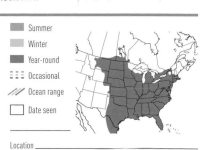

Summer
Winter
Year-round
Occasional
Ocean range
☐ Date seen

____

Location ____________

LOOK FOR: A medium-sized long-legged owl that is brown overall with a white throat and white eyebrows. It has white spots on its back and a brown-striped breast. When open, its large yellow eyes stand out. Stands near burrow or on a low perch, often bobbing its head.

LISTEN FOR: A mellow, dovelike *coo-HOOO* that does not sound like an owl. Also rattles like a rattlesnake from within its nest burrow.

REMEMBER: The Burrowing Owl is the owl most likely to be seen during daylight in open grassland.

WOW!
Home stinky home! Burrowing Owls sometimes line their nest burrows with cowpies!

◀ *Burrowing Owls will readily use artificial nest burrows in appropriate habitat.*

FIND IT: Found in prairies, farmland, and open grassy areas such as airfields, where it nests in underground burrows. Declining because of loss of habitat throughout its range.

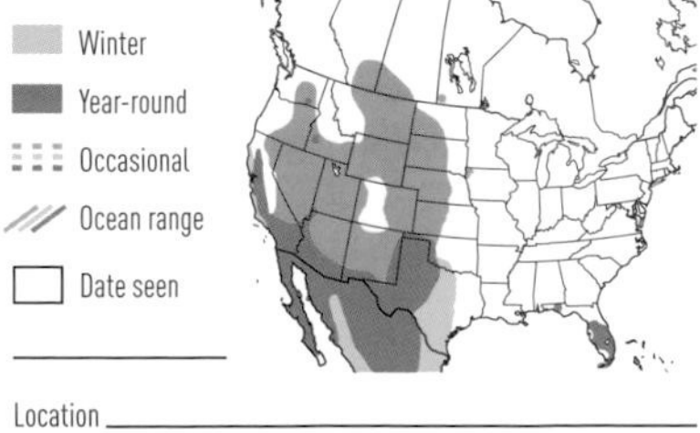

Date seen ______

Location ______

NORTHERN PYGMY-OWL

Glaucidium gnoma Length: 6¾–7"

LOOK FOR: This sparrow-sized owl is round headed and long tailed, with a spotted crown and boldly streaked belly. It bobs its head and flicks its tail while perched. Pursues and catches small birds in flight.

LISTEN FOR: A series of toots on one tone, repeated over and over. Bird watchers in the West often imitate the Northern Pygmy-Owl's tooting call to lure chickadees and other birds into view.

REMEMBER: The Northern Pygmy-Owl is longer tailed than our other small owls.

WOW!
The Northern Pygmy-Owl has fake "eyes" on the back of its head.

◀ *Songbirds, often led by chickadees, will eagerly mob a Northern Pygmy-Owl because this tiny owl eats lots of songbirds.*

FIND IT: Found in a variety of open woods and wooded canyons. Avoids deep woods. Active during daylight, when it perches on interior tree branches, watching for unwary small birds. Responds to imitations of its call.

Summer

Winter

Year-round

Occasional

Ocean range

☐ Date seen

____

Location ______________________________

COMMON NIGHTHAWK

Chordeiles minor Length: 9½" Wingspan: 24"

Adult

LOOK FOR: Overhead in the summer sky, the Common Nighthawk flies with its choppy wingbeats, gliding and swooping after flying insects. Up close, the nighthawk is perfectly camouflaged for daytime perching on the ground or lengthwise along a tree branch. In flight, the bird appears dark overall with a white slash across each wing.

LISTEN FOR: Flying nighthawks utter a sharp, nasal call: *beeertt!* Males performing their courtship flight dive from a great height, making a deep, booming noise with their wings.

REMEMBER: Nighthawks are not really hawks at all, though they look hawklike in flight. Scan the sky at dusk for their distinctive shape and flight style.

WOW!

Nighthawks are members of the goatsucker family, which got its name from the mistaken impression that these birds sucked milk or blood from goats and other livestock.

▶ *Common Nighthawks in a flock during fall migration.*

FIND IT: Common in summer, flying night or day over cities, fields, and parks, the Common Nighthawk is often heard before it is seen. In August and September, look for large flocks of migrating nighthawks at dusk.

Summer
Winter
Year-round
Occasional
Ocean range
Date seen ______

Location ______

LOOK FOR: Its cryptic gray-brown coloration lets the Common Poorwill blend in perfectly when roosting on the ground during the day. This smallest of our nightjars has a white throat, large head, and short tail.

LISTEN FOR: Call is a loud, ringing, repetitive *puh-REE-wah* or, if you use your imagination, *poor-WILL.*

REMEMBER: Like other nightjars, the Common Poorwill is heard more often than seen. If you *do* see one, it will appear much smaller than a Whip-poor-will, with rounder wings and a shorter tail.

◀ *Large rictal bristles help the Common Poorwill catch its flying-insect food.*

WOW!

The Common Poorwill was the first bird discovered to hibernate! It can reduce its metabolic rate and body temperature, going into a trancelike torpor for days or even weeks.

FIND IT: Locally common in areas with dry, rocky ground and scattered brush. Roosts by day, hunts from ground at night, flying up to catch insects. Often hunts along roadways where headlights or a flashlight will cause its eyes to reflect bright orange.

Summer
Winter
Year-round
Occasional
Ocean range
Date seen ______

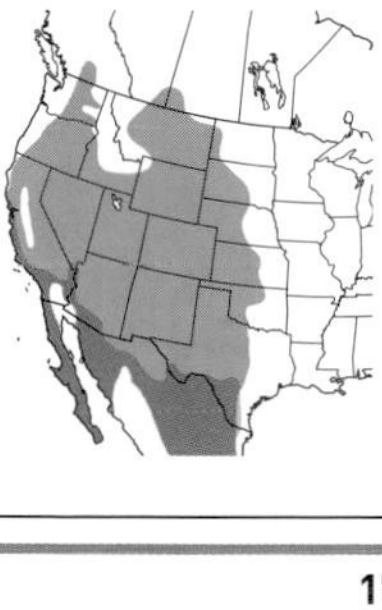

Location ______

WHIP-POOR-WILL

Caprimulgus vociferus Length: 10"

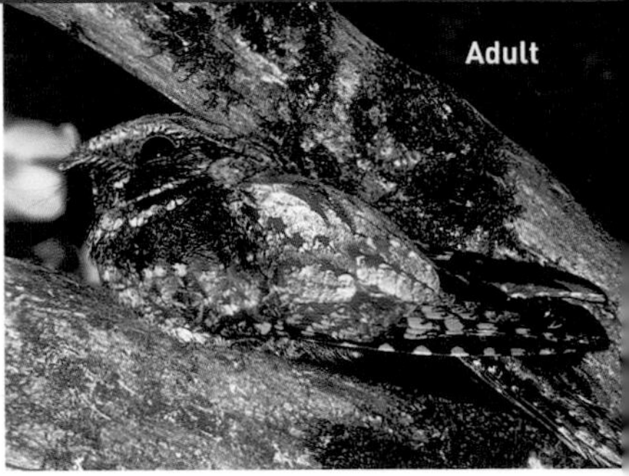

LOOK FOR: The Whip-poor-will is heard far more often than it is seen, but you might catch a glimpse of one as it makes short, wheeling flights after a flying insect. In flight, it is silent and compact-looking with rounded wings. The male Whip shows a flash of white on the outer edges of the tail.

LISTEN FOR: Calls its name over and over again: *whip-poor-WILL, whip-poor-WILL, whip-poor-WILL.* Calls only at night, and only from spring through early fall.

REMEMBER: In the Southeast, the Whip-poor-will is replaced by the larger and browner Chuck-will's-widow. The Chuck also says its name. Voice, not visual clues, is the best way to separate these two closely related species in the area where their ranges overlap.

WOW!

Whip-poor-wills have been timed giving nearly one call per second for more than 15 minutes! That's 1,000 calls in a row without stopping. Don't try this at home!

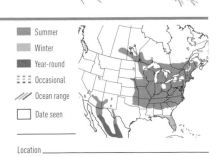

▶ *The Whip-poor-will's large, dark eyes gather light, allowing it to see during its nighttime foraging flights.*

FIND IT: Whip-poor-wills fill eastern woodland nights with their loud, ringing call. Listen for them to start calling at dusk. During the day, Whips roost on low branches or on the ground, relying on their cryptic coloration for camouflage.

Summer
Winter
Year-round
Occasional
Ocean range
Date seen ____________

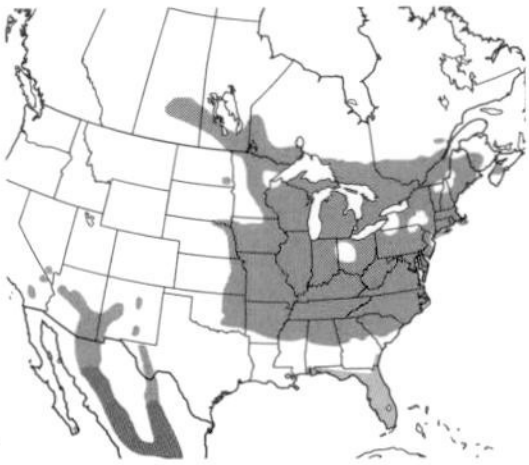

Location ____________

RUBY-THROATED HUMMINGBIRD

Archilochus colubris Length: 3¾"

LOOK FOR: The male's red throat (or gorget) shines brightest when in direct sunlight. At other times it can appear black. The female has a white throat, and both males and females have metallic green backs and wings.

WOW!

Ruby-throats are tiny (weighing less than a penny!) but powerful birds. Their wings flap as fast as 75 beats per second! They are strong fliers, able to migrate 500 miles in one night.

LISTEN FOR: Ruby-throats utter a series of high-pitched twitters almost constantly. They are named for the humming sound made by their wings.

REMEMBER: The tiny Ruby-throated Hummingbird zips past so fast it can be mistaken for an insect, especially a sphinx moth.

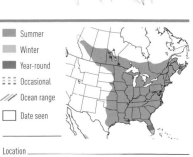

▶ *A male Ruby-throated Hummingbird flashes his red throat in a courtship flight for a perched female.*

FIND IT: The only common hummingbird in the eastern U.S., the Ruby-throat prefers mixed habitat where nectar-producing plants can be found, including parks, gardens, meadows, and woodland edges.

Summer
Winter
Year-round
Occasional
Ocean range
Date seen

____

Location ___________

COSTA'S HUMMINGBIRD

Calypte costae Length: 3½"

LOOK FOR: The male Costa's Hummingbird has a striking, deep purple crown and a gorget that extends downward like a Fu Manchu mustache—unique among our hummingbirds. The female is green above and white below.

LISTEN FOR: Call is diagnostic: a high-pitched, *tinny tink-tink-tink-tink*. Male's display call is a rising, then falling *ziiiing!*

REMEMBER: This species is similar to the Black-chinned Hummingbird, which is slightly larger. The male Black-chinned lacks the extended gorget and purple crown of the male Costa's.

WOW!

This species nests in late winter/early spring in the desert, then migrates west to the Pacific Coast to avoid the desert's summer heat.

▶ *The Costa's Hummingbird is a desert-dwelling bird adept at feeding on desert flowers.*

FIND IT: Common in low-desert habitats: desert scrub, chaparral, dry washes with flowering plants, backyards, and gardens.

____

Location ______________

BLACK-CHINNED HUMMINGBIRD

Archilochus alexandri Length: 3¾"

LOOK FOR: The male's gorget flashes deep purple in the right light; in poor light it can appear black. Both sexes are bright green on the head, back, and tail. Male appears dark headed and shows a clean white bib below the dark purple gorget. Female looks similar to female Ruby-throated Hummingbird.

LISTEN FOR: The wings make a humming sound in flight. Chasing birds give a series of twitters and buzzes. Call is a soft *chew!*

REMEMBER: Black-chinned Hummingbirds pump their tails consistently while hovering.

▶ *Black-chins can be avid feeder visitors.*

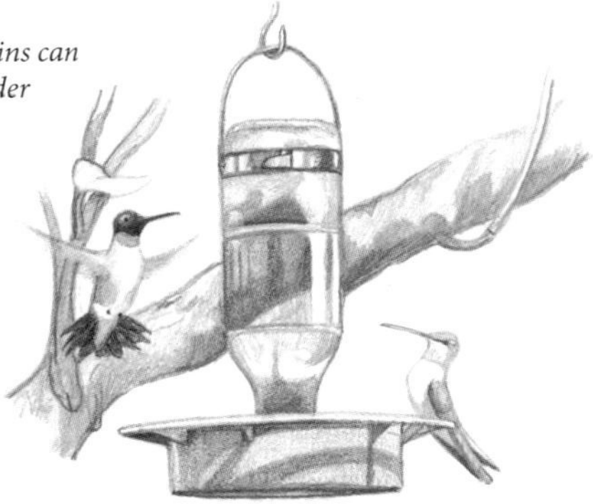

WOW!

In late summer, Black-chinned Hummingbirds move to higher elevations in the mountains to take advantage of blooming flowers.

FIND IT: This is the most widespread hummingbird in the West during summer. Common in wooded foothills and canyons and along the ocean coast. Largely absent from the West in winter.

Summer

Winter

Year-round

Occasional

Ocean range

Date seen

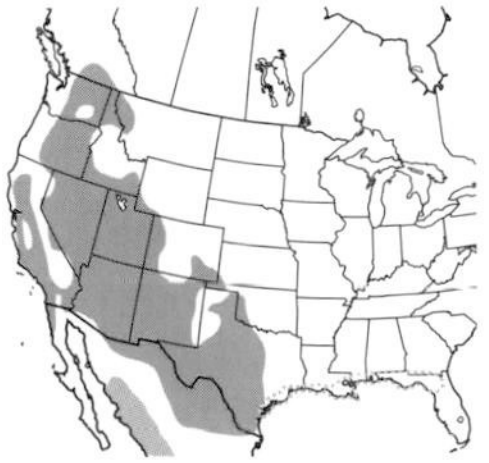

____

Location ____________________

ANNA'S HUMMINGBIRD

Calypte anna Length: 4"

LOOK FOR: The male Anna's Hummingbird is our only hummer with a bright red crown. His crown and gorget are brilliant ruby in good light but can appear black in bright sunshine. A greenish back and smudgy green belly make the Anna's appear "messier" than other hummers. The female may have some red spots on the otherwise white throat.

LISTEN FOR: One of our most vocal hummingbirds. Gives a series of raspy notes and chattery buzzes in what passes for its song, which the male usually delivers while perched.

REMEMBER: At 4 inches long, the Anna's is larger than most other widespread western hummingbirds.

WOW!

Male Anna's can erect gorget and head feathers and move back and forth to flash bright colors at intruders, rivals, and potential mates.

▶ *Male Anna's hummers are avid, if not gifted, singers, usually singing from a perch in their territory.*

FIND IT: The Anna's is common along the Pacific Coast all year long (often the only hummingbird present in midwinter) and is found in brushy habitat and open woods with nectar-producing flowers, including parks, gardens, and backyards.

Summer

Winter

Year-round

Occasional

Ocean range

Date seen

Location ____________________

BROAD-TAILED HUMMINGBIRD

Selasphorus platycercus Length: 4"

WOW!

Broad-tails arriving on territory in early spring before flowers are blooming survive on sap at sapsucker wells.

LOOK FOR: Males and females are metallic green above with a white breast and belly. Male has a magenta gorget. Female has a speckled throat and buffy sides. This hummingbird is often heard before it is seen because the wings of the adult male produce a loud, whistled trill as he flies.

LISTEN FOR: In addition to the wing trill of the male in flight, Broad-tails produce a high-pitched chattering song, usually from a perch. Call is a loud *chit!*

REMEMBER: This species is the most commonly encountered hummer in the Rocky Mountains.

▲ *When the western mountain meadows are full of blooming flowers in summer, Broad-tailed Hummingbirds are easy to find.*

FIND IT: The species is common throughout the inland mountain West, preferring mountain forests, meadows, and wooded canyons.

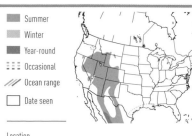

Summer
Winter
Year-round
Occasional
Ocean range
Date seen

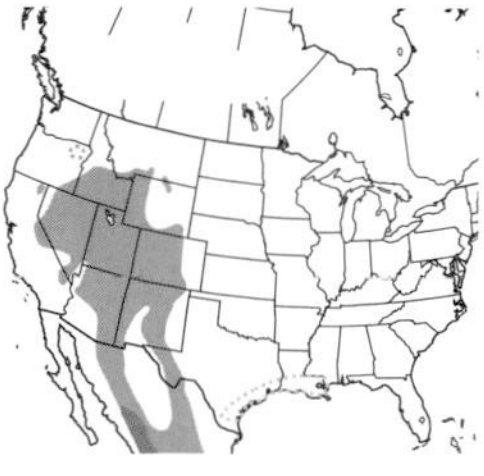

Location ____________

RUFOUS and ALLEN'S HUMMINGBIRDS

Selasphorus rufus, Selasphorus sasin Length: 3¾"

LOOK FOR: These two small hummingbirds are very similar, and females are nearly impossible to tell apart. The male Rufous lives up to its name in color: bright rufous-orange overall, including the back. The male Allen's has a green back and orange tail.

LISTEN FOR: Both vocal and nonvocal sounds of the two species are similar. Aggressive flight call: *zee chippity chippity*. Call note: *chuk*. Wings of males in flight make a high-pitched trill.

REMEMBER: Rufous: none of our other hummers has a rufous back. Allen's: looks like a Rufous Hummingbird but with a *green* back.

WOW!

Individuals of both species can be found wintering in the eastern and southeastern U.S. but the Rufous is the more likely one to see.

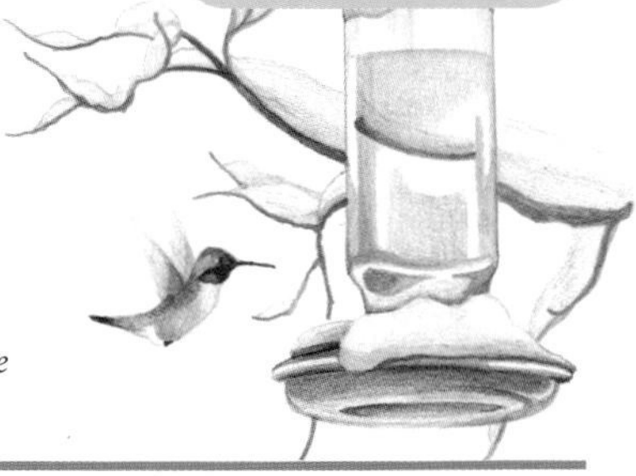

▶ *Rufous Hummingbirds breed as far north as Alaska, where late-spring snows can make finding flower nectar more difficult.*

FIND IT: During the spring and summer: wooded, brushy areas, canyons, open forest, parks, gardens, backyards. In migration both species seek out mountain meadows with flowering plants. Breeding range of Allen's is mostly within California. Rufous is more widespread.

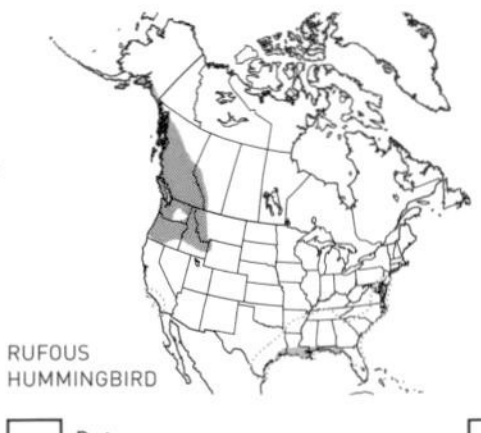

☐ Date seen ____________ ☐ Date seen ____________

Location ____________ ____________

CALLIOPE HUMMINGBIRD

Stellula calliope Length: 3¼"

LOOK FOR: A very small short-billed hummingbird that's shiny green above and white below. Adult male has a gorget that appears striped with magenta feathers. Female is plainer overall with a white spot at the base of the bill.

LISTEN FOR: Song is a high whistle. Calls consist of high-pitched chips and buzzes in a series.

REMEMBER: Besides its small size, another good field mark is this: when perched, the Calliope Hummingbird has wingtips that extend past the tip of the tail.

► *Male Calliopes perform a striking U-shaped courtship flight to impress prospective mates.*

WOW!

The Calliope is our smallest North American bird species, but it can survive the very cold nights in the Rocky Mountains where it nests in the summer.

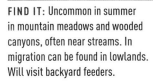

FIND IT: Uncommon in summer in mountain meadows and wooded canyons, often near streams. In migration can be found in lowlands. Will visit backyard feeders.

Summer
Winter
Year-round
Occasional
Ocean range
Date seen

____

Location ______________

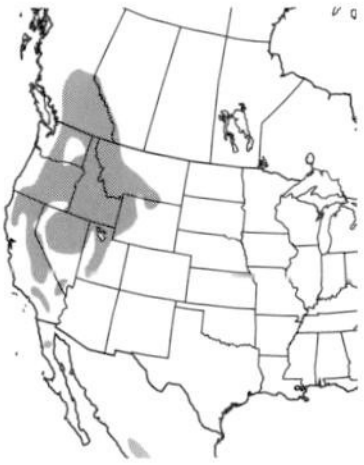

CHIMNEY SWIFT

Chaetura pelagica Length: 5¼" Wingspan: 14"

Roosting

LOOK FOR: Look! Up in the sky! It's a flying cigar! That's what the flight shape of the Chimney Swift looks like: a cigar-shaped body with long slender wings. Flying style is rapid flapping, with short glides in between. Adults have pale throats, but overall the Chimney Swift looks sooty gray. In early fall, huge flocks of swifts may gather at a large roost site, swirling above it before dropping in for the night.

WOW!

Based on a ratio of body size to wing size, Chimney Swifts have the longest wings of any bird.

LISTEN FOR: As they fly, Chimney Swifts emit a high, sputtering chatter that wavers in pitch.

REMEMBER: Swifts are faster, more jerky fliers than swallows, and compared to swallows, the Chimney Swift appears stub tailed.

▶ *A Chimney Swift pair performing a courtship flight, gliding with wings held in a V.*

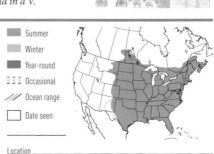

FIND IT: Most common in cities, where they use their namesake chimneys and other hollow structures for nesting and roosting. The Chimney Swift is the only eastern swift species.

Summer

Winter

Year-round

Occasional

Ocean range

Date seen

Location

WHITE-THROATED SWIFT

Aeronautes saxatalis Length: 6½"

LOOK FOR: Our only North American swift species with contrasting black-and-white body, the White-throated Swift usually forages in flocks, flying rapidly and chattering loudly. Throat, central belly, and sides of rump show obvious white.

LISTEN FOR: Loud, chittering *jejejejejeje* dropping in pitch.

REMEMBER: Our other swifts may show pale areas on the throat, but only one of our swifts is truly white throated.

WOW!

A fling on the wing. White-throated Swifts sometimes mate in flight: a male and female meet in the sky, then tumble downward before parting and regaining flight.

► *We know very little about the nesting habits of this species because they nest in such inaccessible places.*

FIND IT: Look up anywhere near a rocky cliff in the West, and these birds are likely to be present. Most common near nesting sites: cliff faces, arid mountains, canyons. Nearly always seen in flight, calling constantly.

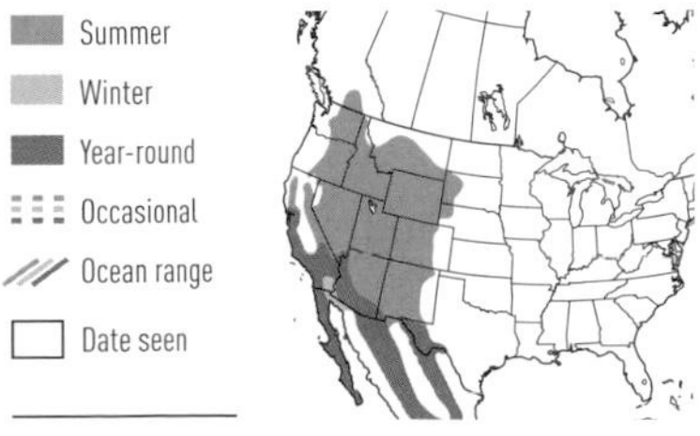

Location ______

BELTED KINGFISHER

Ceryle alcyon Length: 13"

LOOK FOR: This big-headed, shaggy-crested bird is hard to confuse with anything else. The long, daggerlike bill is used to spear fish when the kingfisher plunges from the air into the water. Both sexes have a gray breast-band, but females also have a rusty band (like a bra) across the upper belly.

WOW!

Kingfishers dig nesting holes in vertical soil or sandbanks, often along streams. These nesting tunnels can be up to six feet long!

LISTEN FOR: Most common call is a loud, dry rattle, often given in flight: *ptptptrrrrrrr!* This call has been compared to that of a Hairy Woodpecker.

REMEMBER: In flight, the Belted Kingfisher has jerky, irregular wingbeats. Its white underwings and belly flash as it flies. These clues make it fairly easy to identify a Belted Kingfisher at great distances.

▼ *With beaks and feet, Belted Kingfishers excavate a deep burrow in a sandy bank, where they raise their young.*

FIND IT: A conspicuous bird that perches in the open along streams, rivers, and lakes, the Belted Kingfisher gives a loud, rattling call as it flies. It hovers over the water, making a spectacular plunging dive when it spots a fish.

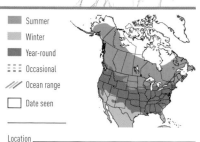

Summer

Winter

Year-round

Occasional

Ocean range

Date seen ______

Location ______

LEWIS'S WOODPECKER

Melanerpes lewis Length: 10¾–11"

LOOK FOR: A large dark woodpecker. Adults have a red face, pinkish belly, dark green back and wings, and silver gray collar. Juveniles have an all-brown face and lack the collar.

LISTEN FOR: Not very vocal. Occasionally utters a high-pitched, sputtery chatter. Drumming is very rapid.

REMEMBER: Unlike the undulating flight of other woodpeckers, the Lewis's flight is even, like a crow's.

Adult

▼ *Lewis's Woodpeckers are experts at flycatching: flying out from a perch to nab a passing insect.*

WOW!
The Lewis's Woodpecker is one of the species discovered and named during the Lewis and Clark expedition to the West in the early 1800s.

FIND IT: Uncommon in summer in open woods, orchards, riparian groves, and in logged or burned woodlands. In summer, open habitat is better for flycatching. Winters in loose flocks near a reliable source of food—such as acorns—in woods of oaks or other nut-bearing trees.

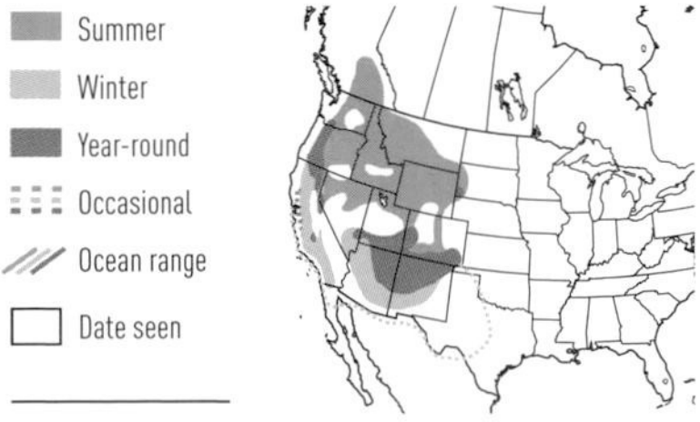

Location ______

Melanerpes formicivorus Length: 9"

LOOK FOR: Loud, active, and boldly marked, the Acorn Woodpecker is hard to miss. Both males and females are black backed and have a facial pattern resembling clown makeup (black and yellow face, white eye, black bill, red head patch), but male has a more extensive red crown. White rump and wing patches stand out in flight.

LISTEN FOR: A very vocal bird. Call is a raucous *wake-up! wake-up! wake-up!* Also sputtering *churrs.*

REMEMBER: Other black-backed woodpeckers of the West lack the Acorn's clown makeup.

WOW!

One sycamore tree in California used by Acorn Woodpeckers as an acorn "granary" held an estimated 20,000 acorns! And this did not kill the tree!

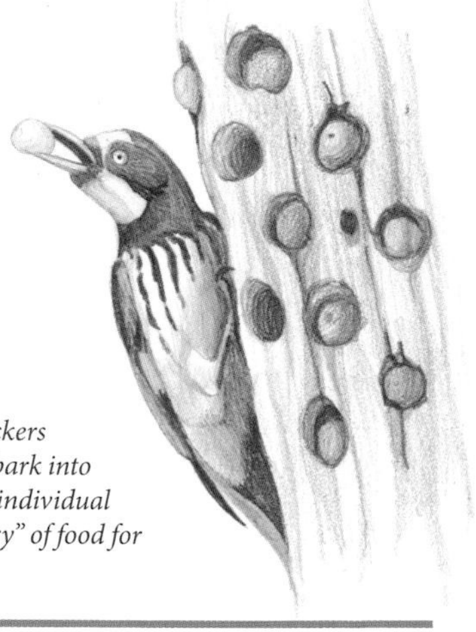

▶ *Acorn Woodpeckers drill holes in tree bark into which they insert individual acorns as a "pantry" of food for later in the year.*

FIND IT: As the name suggests, the Acorn Woodpecker usually associates with acorn-bearing oaks of several species. Lives in small, very social, and conspicuous colonies in mixed oak-pine woods, canyons, foothills, and in wooded residential areas.

Summer
Winter
Year-round
Occasional
Ocean range
Date seen ________

Location ________

LOOK FOR: The adult Red-headed Woodpecker has an all-red head and a body that goes black-white-black from back to tail. In flight, the white wing patches flash and are an excellent field mark on distant birds. Young Red-heads have brown heads until their first full spring.

LISTEN FOR: Call is an excited-sounding *queerp!* Also gives a *chrrr* call in flight. Vocalizations are similar to those of the Red-bellied Woodpecker, but the Red-headed sounds more excited.

REMEMBER: Only the Red-head has an *all-red head.* The male Red-bel lied Woodpecker has a mohawk of red across the top of the head from front to back.

WOW!

Red-headed Woodpeckers are excellent fliers, and they use their aerial skills to catch flying insects.

◀ *Most battles between Red-headed Woodpeckers and European Starlings occur over nest cavities. The woodpecker usually wins.*

FIND IT: Locally common in stands of trees or open woods, the Red-head is easy to see when it flies. When perched, though, it often sits quietly and may be overlooked. It is loosely colonial, so where you find one bird, there may be others.

Summer
Winter
Year-round
Occasional
Ocean range
Date seen

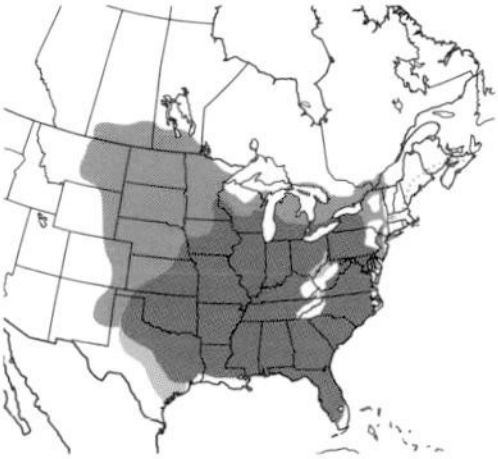

____

Location ______________

RED-BELLIED WOODPECKER

Melanerpes carolinus Length: 9¼"

LOOK FOR: All Red-bellied Woodpeckers have zebra-striped backs; the male has a band of red on the head from the bill to the nape. Females have red napes only—their crowns are brown. Flight is strong and swooping and shows flashes of white in the wings and tail.

LISTEN FOR: Red-bellies make a variety of sounds, including a short call that rises in tone: *quiirrr!* And a longer call: *ch-ch-ch-chirrrrrrrr.* Avid drummers, they sometimes use metal chimneys and drain-pipes for maximum noise.

WOW!

This species will eat almost anything, including insects, small fish, tree frogs, and even other birds and bird eggs!

REMEMBER: The Red-bellied Woodpecker is named for a field mark that is very hard to see. Any red on the belly is hidden from view as the bird is perched against a tree trunk or branch.

▶ *Red-bellied Woodpeckers often rule the feeder. Here, one threatens a White-breasted Nuthatch.*

FIND IT: This species will eat almost anything, including insects, small fish, tree frogs, and even other birds and bird eggs!

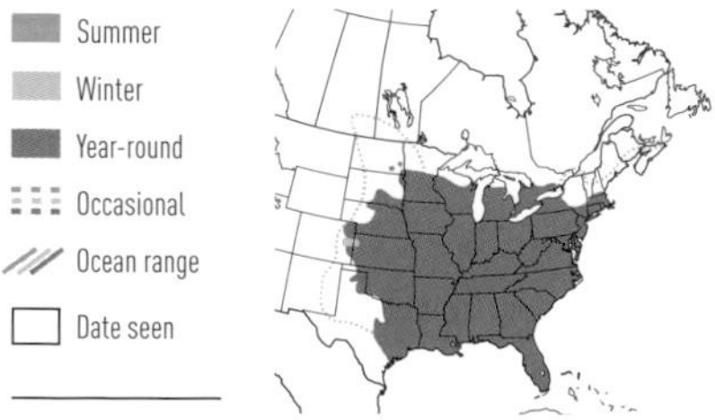

Location ______

Melanerpes uropygialis Length: 9¼"

LOOK FOR: A tan-bodied medium-sized woodpecker with a zebra-striped back and wings and a plain face. Male has a round red cap (female lacks this). Dark eye and bill stand out on tan head.

LISTEN FOR: Call is a burry, rising *churr!* Also gives a series of squeaky calls: *earp! earp! earp!* Drumming is loud but spaced out and slows near the end.

REMEMBER: Farther to the east, in Texas and Oklahoma, lives the very similar Golden-fronted Woodpecker, the male of which has a golden forehead and nape. Similar Ladder-backed Woodpecker has a striped face.

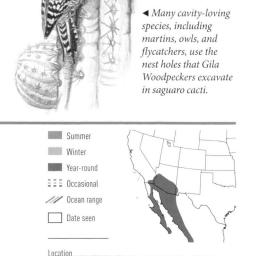

◀ *Many cavity-loving species, including martins, owls, and flycatchers, use the nest holes that Gila Woodpeckers excavate in saguaro cacti.*

WOW!

Gila Woodpeckers will eat almost anything, including the eggs and young of other birds! But they are otherwise very nice birds.

FIND IT: A woodpecker of the low southwestern deserts, where it prefers desert washes, riparian woodlands, and towns, especially in areas with saguaro cacti.

Summer

Winter

Year-round

Occasional

Ocean range

☐ Date seen

____

Location ______________

WOW!

Ornithologists have changed the poor flicker's name many times over the years, from Yellow-shafted, to Common, back to Yellow-shafted, then to Northern Flicker.

LOOK FOR: The Northern Flicker occurs in two distinct forms: Yellow-shafted (a mostly eastern bird) and Red-shafted (mostly western), named for the color on the undersides of the wings. Other field marks include a black breast-band, black belly spot, red nape spot, and a black mustache (on Yellow-shafted males).

LISTEN FOR: The flicker's main call is a high-pitched, explosive *kleer!* Or a long, even series of maniacal-sounding notes: *bir-bir-bir-bir-bir.* A nasal *wik-a-wik-a-wik-a-wik* call accompanies courtship and territorial displays. Frequently drums loudly.

REMEMBER: The flash of color in its wing and the brilliant white rump can ID a flicker even at a great distance.

▼ *A Northern Flicker eats plant seeds in winter when insects are not available.*

FIND IT: Prefers open areas of scattered trees, such as parks, cemeteries, and backyards, over deep woods. Of all of our woodpeckers, this is the one most likely to be seen on the ground, where the flicker loves to eat ants.

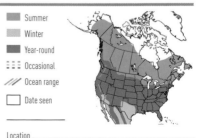

Summer

Winter

Year-round

Occasional

Ocean range

Date seen

Location

RED-BREASTED SAPSUCKER

Sphyrapicus ruber Length: 8½"

LOOK FOR: Of our four sapsucker species, only the Red-breasted has an all-red head and breast. The back and tail are loosely zebra striped, and the black wings have a bold white stripe that is vertical on perched birds. Adults are similar. Juvenile birds are dusky headed.

LISTEN FOR: Loud, descending *cherrr!* nasal in tone. Also a wild-sounding series of squeals. Drumming is irregular clusters of beats: *ratatat-tatat-rata-tatatat.*

REMEMBER: This is one of the few woodpecker species in which the adult males and females look alike.

WOW!

Once considered part of the same species as the Yellow-bellied Sapsucker, the Red-breasted will interbreed with the Yellow-bellied and the Red-naped in the limited areas where their ranges overlap.

◀ *Sapsuckers will chase away other birds and animals that try to mooch a meal at their sap wells. This moocher is a Rufous Hummingbird.*

FIND IT: Common in mixed deciduous-coniferous woods and groves along the western edge of North America. This sapsucker species is less migratory than its closely related species.

Summer
Winter
Year-round
Occasional
Ocean range
Date seen ______

Location ______

YELLOW-BELLIED SAPSUCKER

Sphyrapicus varius Length: 8½"

LOOK FOR: The yellow belly for which this bird is named is visible only on flying adult birds. Adults have red crowns, and males add a red throat to the mix. The best field mark is the long vertical white slash along the wing. First-year birds have mottled brown heads and breasts.

LISTEN FOR: Usually silent in winter but vocal on the breeding grounds and in spring migration, when it utters a wheezy, catlike *meeyaah!* Light tapping as it drills sap holes can be a good way to locate this species.

REMEMBER: This is our only eastern sapsucker, but there are three closely related sapsucker species in the West: the Red-breasted, Red-naped, and Williamson's Sapsuckers.

▼ *An adult male Yellow-bellied Sapsucker defends its sap wells from a curious Ruby-throated Hummingbird.*

WOW!

It's no joke! The Yellow-bellied Sapsucker has a silly name. The truth is they do eat sap, and their sap holes are also used by many other birds, mammals, and insects as sources of food.

FIND IT: Yellow-bellied Sapsuckers are especially fond of fruit trees and other trees that readily ooze protective sap. They can be creatures of habit, returning to visit old sap holes year after year. Look for their rings of sap holes on tree trunks.

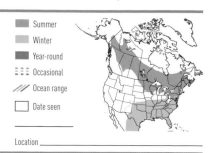

Summer
Winter
Year-round
Occasional
Ocean range
Date seen

______

Location ______

LADDER-BACKED WOODPECKER

Picoides scalaris Length: 7¼"

LOOK FOR: It's hard to miss the horizontal black-and-white "ladder" stripes across the back of the Ladder-backed Woodpecker. The boldly marked face shows a white cheek. Adult male has a red crown.

LISTEN FOR: Common call is a high-pitched *peek!* Also a laughing rattle call, dropping off in pitch at the end. Drumming is a fast drumroll.

REMEMBER: A woodpecker with a zebra-striped back in arid southwestern habitat is likely to be this species.

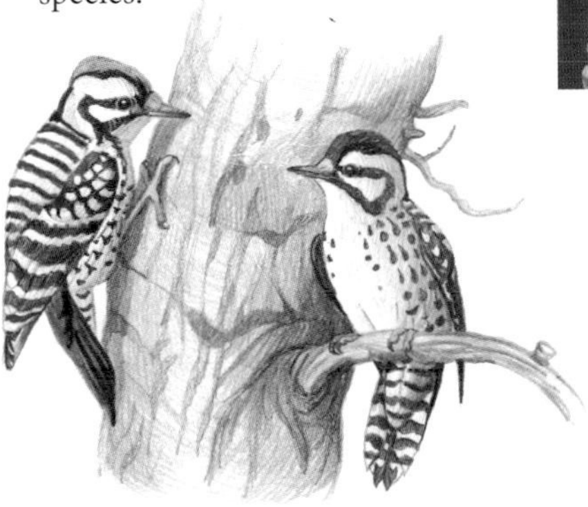

▲ *Mated pairs of Ladder-backs forage together but concentrate on different parts of the tree: males on larger limbs and trunks, females on smaller ones.*

WOW!

Two of the folk names for the Ladder-backed Woodpecker are Cactus Woodpecker (for its preferred nest site) and Mexican Woodpecker (for its range). Neither one of these is as descriptive as Ladder-backed.

FIND IT: A permanent resident in the dry mesquite and wooded scrub habitats of the Southwest, the Ladder-backed Woodpecker fills the same ecological niche as the Downy Woodpecker, which does not occur in this specific habitat.

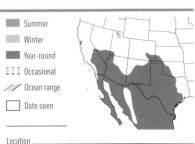

Summer
Winter
Year-round
Occasional
Ocean range
Date seen

____

Location ______________

Picoides villosus Length: 9"

LOOK FOR: The Hairy Woodpecker looks like a supersized version of the Downy Woodpecker. The best field mark for separating the two is bill length. The Hairy has a large bill that's as long as its head is wide.

LISTEN FOR: Hairy Woodpeckers even *sound* bigger than Downies. Their call note, *peek!*, is sharper and louder than the Downy's, and the Hairy's rattle is lower and more emphatic and *does not drop in pitch*. They also drum on hollow branches.

REMEMBER: Despite their larger size, Hairy Woodpeckers can be shier than Downies. When a Hairy sees you approaching, it may scoot around to the back of the tree trunk it's on before flying off.

▶ *Downy (right) and Hairy Woodpeckers on a suet feeder. The Hairy's larger body, head, and bill are easy to see.*

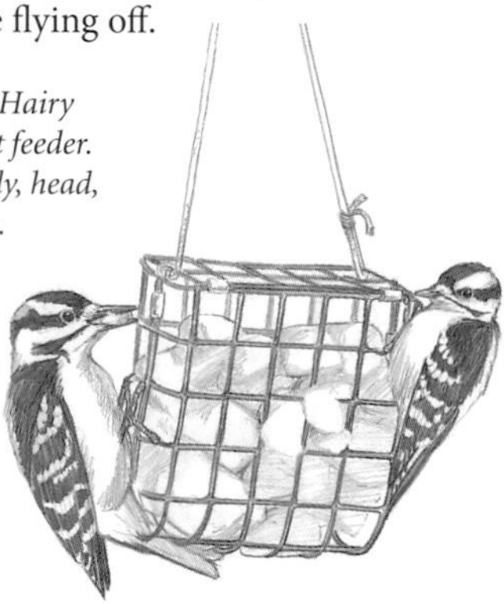

WOW!

Some woodpeckers attack house siding. They may be able to hear wood-boring insects inside a tree trunk but mistake the hum of electricity in household wiring for insect activity.

FIND IT: Hairies can be found year-round across the continent, but they need larger trees than Downies do for foraging and nesting, so they are not as common in open areas. At bird feeders, Hairies eat suet and peanuts.

Summer
Winter
Year-round
Occasional
Ocean range
☐ Date seen

____

Location ________________________________

LOOK FOR: The Downy is the smaller of our two common black-and-white woodpeckers (and North America's smallest woodpecker). Its bill is short, equal to about half of the width of its head. The male Downy has a small red spot on the back of its head (females have no red).

LISTEN FOR: Downies give a high-pitched *pik!* call as well as a long, ringing rattle, *trrrrrrrrrrr!*, that descends in pitch (the Downy's rattle goes *down*). Also drums on resonating hollow branches.

REMEMBER: Downy is "dinky," Hairy is "huge." Telling these two species apart is easy once you remember this size difference. The Hairy's bill is as long as its head is wide.

WOW!

Every fall, male and female Downies each excavate their own roost holes. These are the cozy places where they spend the chilly winter nights.

▶ *The Downy Woodpecker's winter diet includes larvae it extracts from galls on goldenrod stems.*

FIND IT: Widespread and common wherever there are trees for foraging and nesting. Downies often perch on weed stalks and do not require the large trees needed by the Hairy Woodpecker. Frequents backyard bird feeders.

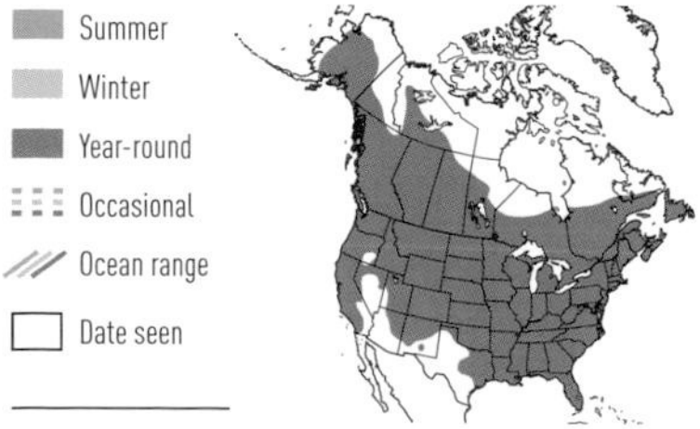

Location ____________

PILEATED WOODPECKER

Dryocopus pileatus Length: 17"

WOW!

If you see a large rectangular hole chiseled in a large tree, it's the work of the Pileated Woodpecker. They need to create large holes to get to ants and grubs in the rotting centers of old trees.

LOOK FOR: The Pileated is difficult to confuse with any other bird because of its large size and obvious red crest. Perched, the Pileated appears mostly black except for the head and neck. The bright white underwings are an excellent field mark on distant flying Pileateds.

LISTEN FOR: Pileateds are very vocal; the series of notes in its long, loud call sounds like someone laughing. It is similar to the call of the Northern Flicker, but louder and more excited. In flight, they may utter single or double notes of the same quality. Also drums loudly, with a pattern that slows down as it ends.

REMEMBER: The much rarer (and possibly extinct) Ivory-billed Woodpecker has a white bill and black throat, and shows huge white wing patches when perched.

▲ *In fall and winter, Pileated Woodpeckers rely on fruits and berries for much of their diet.*

FIND IT: As the woodlands of North America have returned, so has the Pileated Woodpecker. It is most common in mature woods with large trees but also found in city parks and suburban neighborhoods.

Summer
Winter
Year-round
Occasional
Ocean range
Date seen
____

Location _______________

EASTERN and WESTERN WOOD-PEWEES

Contopus virens, Contopus sordidulus Length: 6½", 6¼"

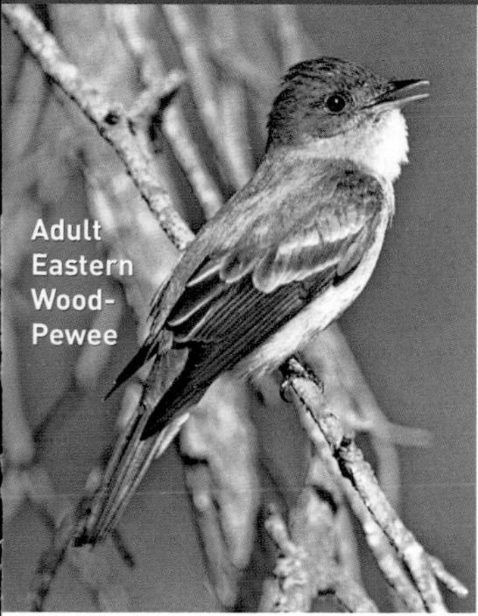

LOOK FOR: The Eastern Wood-Pewee is similar to the Eastern Phoebe but smaller. Two obvious pale wing bars, lack of a distinct eye-ring, and a lack of tail wagging are good ID clues for the pewee. Western: dark "vest" on chest is stronger than that of Eastern, wing bars are less distinct.

LISTEN FOR: The song of the Eastern Wood-Pewee is *pee-ah-WEE!* or *PEE-yerr!* Western: nasal *pee-yerrr*. Pewees often sing throughout the day and continue even after dusk, when many other birds are silent.

REMEMBER: Eastern Wood-Pewees do not wag their tails while perched, but many birders have noted that wood-pewees often flick their wings right after landing on a perch. Tail wagging = Eastern Phoebe. Wing flicking and no tail wagging = Eastern Wood-Pewee.

WOW!

The Eastern Wood-Pewee's western cousin is the Western Wood-Pewee. Fortunately, these two look-alike species have separate ranges and different voices, otherwise we'd never be able to tell them apart.

▼ *Even though it's exposed, the Eastern Wood-Pewee's nest is inconspicuous, resembling a lichen-covered knot on a branch.*

FIND IT: You are likely to hear a wood-pewee before you see it. Look for Easterns in summer along woodland edges, perched high in trees on internal bare branches, from which they make regular foraging flights. Western: prefers pine-oak forests, mature woods, woodland edges.

☐ Date seen ______________ ☐ Date seen ______________

Location ______________ ______________

ACADIAN FLYCATCHER

Empidonax virescens Length: 5¾"

LOOK FOR: Among the small, drab, and confusingly similar *Empidonax* flycatchers, the Acadian Flycatcher is perhaps the most commonly encountered. It has the two pale buff wing bars and eye-ring of its fellow Empids, but its body shows more contrast. Its head often shows a peak at the back. Acadians sing frequently, so learning their call is the best way to identify them.

LISTEN FOR: A high, sharp call that sounds like a squeeze toy: *peet-ZUP!* At dawn and dusk it also gives a longer, more elaborate song that is very sputtery and explosive.

REMEMBER: The Acadian Flycatcher is the "hungry" flycatcher. It orders a "pizza" when it calls: *peet-ZUP!*

WOW!

Acadian Flycatchers spend summers with Wood Thrushes and Ovenbirds, then migrate across the Gulf of Mexico to spend the winter in tropical cloud forest with quetzals and bellbirds.

◀ *The Acadian Flycatcher's nest looks like debris caught in a limb. It is a well-woven basket hung between two twigs.*

FIND IT: In woodlands, especially near water, the Acadian usually perches on a branch halfway up a tree that is inside the forest (rather than on the edge). From there it calls and makes short flights out to capture insects.

Summer
Winter
Year-round
Occasional
Ocean range
Date seen ______

Location ______

LOOK FOR: As its name implies, this is a small flycatcher, olive green above, pale greenish below, with a bold white eye-ring and two obvious white wing bars.

LISTEN FOR: A loud, emphatic *che-bek!* repeated in rapid succession. Call note: *whit!*

REMEMBER: It's really hard to tell the *Empidonax* flycatchers apart. But most ID experts agree that the Least Flycatcher is the smallest and grayest of the bunch.

► *The Least Flycatcher's voice may lack melody, but it's the best ID clue for this species.*

WOW!

Least Flycatchers don't like American Redstarts as neighbors. They regularly chase redstarts (which compete for food) out of their nesting territories.

FIND IT: Common in summer in mature deciduous and mixed woods, where it often perches on a bare branch in the middle of a tree or lower. Often heard before it is seen.

Summer

Winter

Year-round

Occasional

Ocean range

☐ Date seen

____

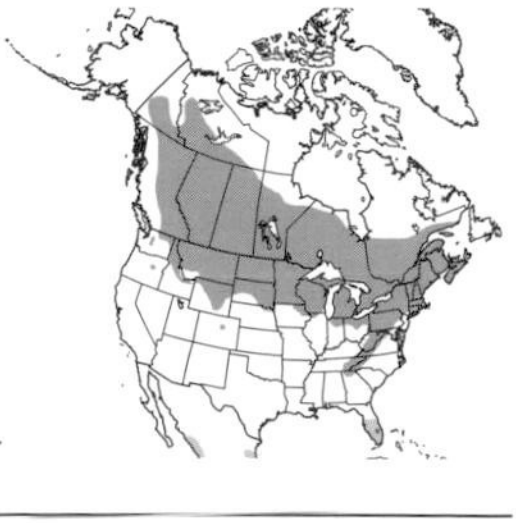

Location ______________________________

WILLOW FLYCATCHER

Empidonax traillii Length: 5¾"

LOOK FOR: Olive green above, dusky pale yellow below with a barely noticeable eye-ring and weak, drab wing bars. Bill is longish and dark above, yellow below. Like other Empids, often heard before it is seen.

LISTEN FOR: The Willow's *FITZ-bew!* call is explosive and sounds like the bird is sneezing.

REMEMBER: Voice is the best way to separate this species from its close relatives.

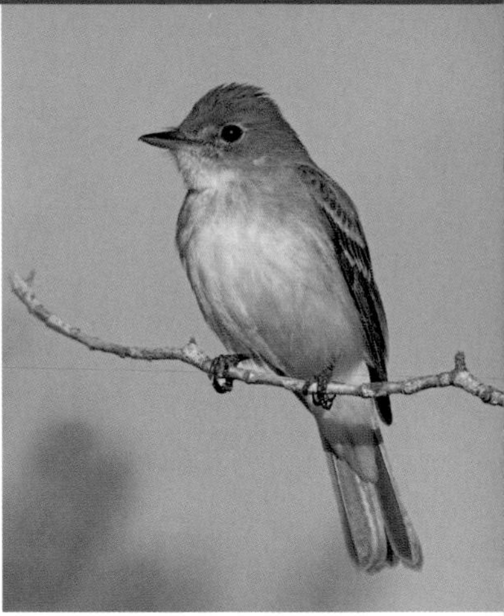

WOW!

The Willow Flycatcher and its close relative the Alder Flycatcher were considered a single species (Traill's Flycatcher) until the 1970s.

◀ *Don't be fooled! Willow Flycatchers are* not always *found in willows. This bird sings from a streamside cottonwood.*

FIND IT: The Willow Flycatcher spends the summer in wet willow thickets, brushy old-fields, and woodland edges. Its distribution is mostly south of the Canada border, while the similar Alder Flycatcher is more common north of the border.

Summer

Winter

Year-round

Occasional

Ocean range

Date seen ______

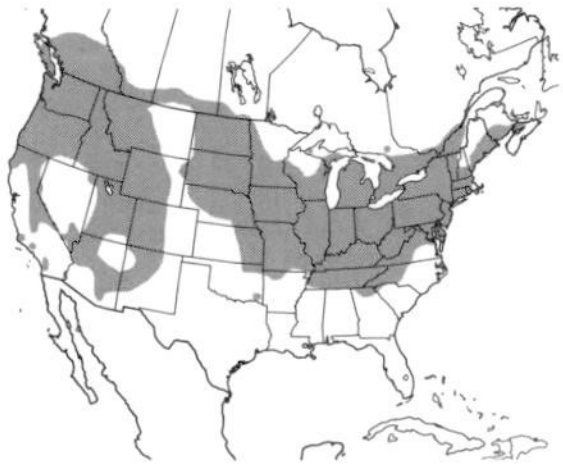

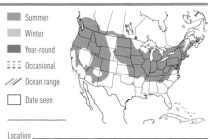

Location ______

Sayornis nigricans Length: 6¾–7"

LOOK FOR: Dark above, white below, the Black Phoebe is our only flycatcher with a black breast. Except for its flycatching behavior and wagging tail, it looks much like a Dark-eyed Junco.

LISTEN FOR: A thin, whistled *fee-bee, fee-bew*. Call note is a *chep!*

REMEMBER: Like other phoebes, the Black Phoebe bobs its tail constantly while perched.

WOW!
Some Black Phoebes have been observed catching and eating small fish!

▶ *Black Phoebes build their nests out of mud and grass and line them with animal hair. The mud helps the nest stick in place.*

FIND IT: Always found near water and a year-round resident in most of its range. Along streams, ponds, and lakes, in towns, and even near farmyard troughs, the Black Phoebe can be found perched low and catching insects just above the water's surface.

Summer
Winter
Year-round
Occasional
Ocean range
Date seen

____

Location ______________

LOOK FOR: The drab-looking Eastern Phoebe is considered by many birders to be a sign of spring's arrival. Calling out its name and flicking its tail up and down, this medium-sized flycatcher returns in spring as soon as insects are active. Phoebes are dark headed and pale bellied. Juvenile birds can show yellow bellies.

LISTEN FOR: Eastern Phoebes call out *FEE-bee!* or *FEE-bree!* They also utter a soft, sweet-sounding chip note.

REMEMBER: Phoebes can be confused with the smaller Eastern Wood-Pewee, but phoebes wag their tails and pewees do not. Pewees always show obvious white wing bars. You have to look really hard to see the phoebe's wing bars.

WOW!

The first North American ornithologist to band birds, John James Audubon tied a small silver thread around the legs of nestling Eastern Phoebes.

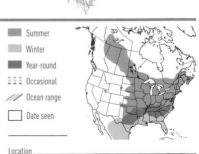

▶ *Phoebes will build their mud and moss nests on any ledge that is protected from the weather.*

FIND IT: Eastern Phoebes prefer wooded habitat near water. They build mud nests under overhanging rocks, under bridges, and in barns and can be reliably found in these settings during warm months.

Summer

Winter

Year-round

Occasional

Ocean range

Date seen ________

Location ________

SAY'S PHOEBE

Sayornis saya Length: 7½"

LOOK FOR: A medium-sized gray-brown flycatcher with rufous pink sides and a square black tail. Dark eye line stands out on gray head. Upperwings, best seen in flight, are plain gray. Like other phoebes, wags tail while perched.

LISTEN FOR: A sad-sounding, down-slurred call: *pee-yerr* or *pyeer*. Sometimes alters the two phrases.

REMEMBER: While our other two phoebes prefer to be near water, Say's Phoebe can survive in much drier habitats.

WOW!

Say's Phoebe has a huge breeding range, from the desert Southwest to the tundra habitats near the Arctic Circle.

▶ *On the windblown prairie of North Dakota, a Say's Phoebe perches near its nest in an abandoned farm building.*

FIND IT: Prefers open habitat, including prairie, scrublands, canyons, ranches, and parks.

Summer

Winter

Year-round

Occasional

Ocean range

Date seen

____

Location ______________

VERMILION FLYCATCHER

Pyrocephalus rubinus Length: 6"

WOW!

The male Vermilion performs an amazing courtship flight above his territory, fluffing out his red feathers while fluttering and singing, then swooping back to a prominent perch.

LOOK FOR: It's hard to misidentify a male Vermilion Flycatcher, with his flaming red crown, throat, and underparts offset by the dark brown mask and upperparts. Adult female is pale gray above and streaky chested with a dark mask and black tail. Some adult females show a pink wash on the belly.

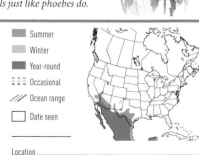

LISTEN FOR: Song is a series of rising, staccato notes, ending in a trill: *pit-pitpit-zree!*

REMEMBER: Females can have pinkish or yellowish wash on lower belly. Brown breast streaks, dark mask, and short dark tail are best field marks.

▸ *When perched, Vermilion Flycatchers wag and flare their tails just like phoebes do.*

FIND IT: Found near open water in dry habitats with scattered trees in the Southwest. Also found along streams, near ponds. Most noticeable as it sorties into the air after insects and returns to a perch.

Summer

Winter

Year-round

Occasional

Ocean range

Date seen

____

Location ________________

GREAT CRESTED FLYCATCHER

Myiarchus crinitus Length: 8½"

LOOK FOR: This tall flycatcher's most obvious field marks are its crested head, gray face and throat, bright lemon yellow belly, and rufous tail, but Great Cresteds are most often located by their voices. Three other close *Myiarchus* relatives occur in the West and Southwest, but the Great Crested is the only one in the East.

LISTEN FOR: A loud, upslurred *whee-eep!*, given singly or in a sputtering series.

REMEMBER: You might confuse this flycatcher with a kingbird, but the Great Crested is far more colorful on its belly and tail.

WOW!

This flycatcher often includes shed snakeskin in its nest—perhaps to deter predators, but they also simply like the crinkly feel of the skin.

▼ *A Great Crested Flycatcher brings food to a hungry nestling.*

FIND IT: Great Cresteds prefer older woods but may also be found in backyards and parks with tall, leafy trees. They nest in tree hollows and old woodpecker holes and may occasionally use nest boxes.

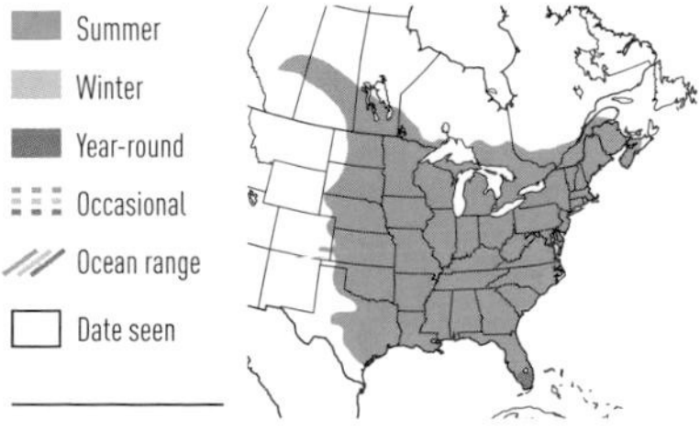

Location ______

WESTERN KINGBIRD

Tyrannus verticalis Length: 8¾"

LOOK FOR: Among our common flycatchers, the Western Kingbird, with its pale gray head, white throat, and lemon yellow belly, stands out. In flight, the black tail shows white edges.

LISTEN FOR: Short squeaky call notes. Also a series of explosive, sputtery squeaks like a squeeze toy getting chewed on by several puppies at once: *pick! peepick! pick! peekaboo!*

REMEMBER: Where their ranges overlap, the Western Kingbird prefers more open habitat than the Eastern Kingbird.

WOW!
Western Kingbirds will boldly attack a crow, raven, or hawk that passes near their nesting territory.

▶ *Roadside fences provide an excellent launching pad for catching flying insects.*

FIND IT: Common in summer throughout the West in open country with scattered trees and along roadside fences in farmyards and towns. Perches in the open, flying out to catch insects in the air.

Summer
Winter
Year-round
Occasional
Ocean range
Date seen ______

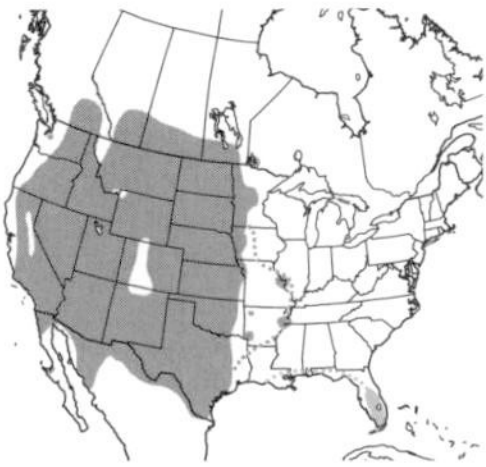

Location ______

EASTERN KINGBIRD

Tyrannus tyrannus Length: 8½"

LOOK FOR: Black above, white below, with a heavy black bill, the Eastern Kingbird looks a bit mean. As its Latin name suggests, it is a tyrant of tyrants, often attacking much larger birds that invade its territory. The obvious white tip to the black tail is a field mark unique to this flycatcher species.

LISTEN FOR: Eastern Kingbirds are loud birds (often heard before seen) with an unmusical, zapping call: *ptzeent!*, often given in a rapid series.

REMEMBER: From the Great Plains westward, you can see both Eastern and Western Kingbirds in the same habitats. They are easy to tell apart—Westerns are light gray above and yellow below.

▼ *Eastern Kingbirds overwintering in the Neotropics travel in flocks and feast on the fruit of Cecropia trees.*

WOW!

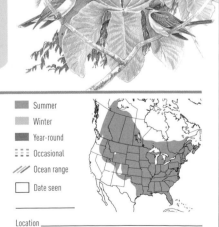

If you get a really good look, you'll see a narrow stripe of red feathers on the crown of an Eastern Kingbird. This rarely seen plumage is displayed by the male during spring courtship.

FIND IT: Common in summer in a variety of open and semi-open habitats across the continent, Eastern Kingbirds perch in obvious places—on roadside wires, fences, and treetops—and make aerial forays to catch insects.

Summer
Winter
Year-round
Occasional
Ocean range
Date seen ______

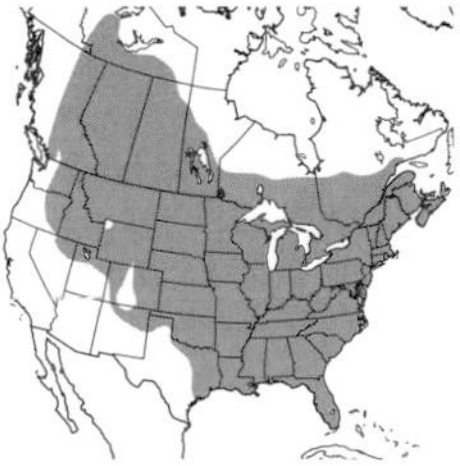

Location ______

SCISSOR-TAILED FLYCATCHER

Tyrannus forficatus Length: 13"

LOOK FOR: Though it is not common in the East, this spectacular pale flycatcher with the elegant, long black-and-white tail is a must-see for every birder. As if the tail were not impressive enough, the Scissor-tailed Flycatcher has a wash of bright pinkish orange on its sides and in its wingpits.

LISTEN FOR: Common call is sputtery and rising in tone: *pik-pik-pik-pik-piDEEK!*

REMEMBER: Not all Scissor-tails have superlong tails. Females and juveniles have shorter tails on average.

WOW!

In spring, the male performs an incredible courtship flight. Climbing 100 feet up, he dives, somersaulting and showing off his tail, then tumbles back to a perch. This never fails to impress the girls.

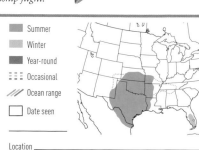

▶ *The male Scissor-tailed Flycatcher looks like a bird of paradise as it makes a swooping courtship flight.*

FIND IT: In summer it perches along fence lines and in treetops, making foraging flights to catch insects. In fall, large migratory flocks are common in Texas. Individuals may be found far beyond the normal range in spring and fall.

Summer
Winter
Year-round
Occasional
Ocean range
Date seen ______

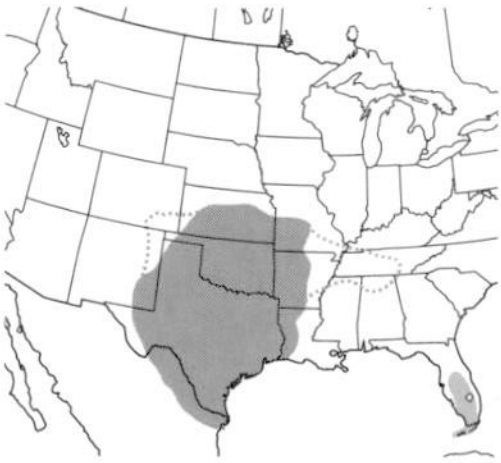

Location ______

LOOK FOR: The Loggerhead Shrike is a songbird with a killer instinct, catching a variety of insects, small mammals, reptiles, and other birds with its hooked, hawklike bill. The stout black bill and black mask give this bird a dark-headed look. In flight, the Loggerhead Shrike shows white patches on black wings and tail, similar to a Northern Mockingbird.

LISTEN FOR: The Loggerhead Shrike has a surprisingly musical voice. Short, burry whistles are interspersed with harsh, nasal notes and buzzes.

REMEMBER: Look at the black facemasks to tell the Loggerhead Shrike from the (less common) Northern Shrike and Northern Mockingbird. Mockingbirds lack a mask. Northern Shrikes have barely any mask. The Loggerhead's mask seems to connect to the black bill.

▶ *The last thing many voles see is a Loggerhead Shrike's masked face.*

WOW!

The Loggerhead Shrike and its relative the Northern Shrike have the folk name of Butcherbird for their habit of impaling prey on thorns or fence wire as a butcher hangs out slabs of meat.

FIND IT: Loggerhead Shrikes prefer exposed perches in open country from which they watch for prey. The Loggerhead is disappearing from most of the Northeast. The cause of this population decline is poorly understood.

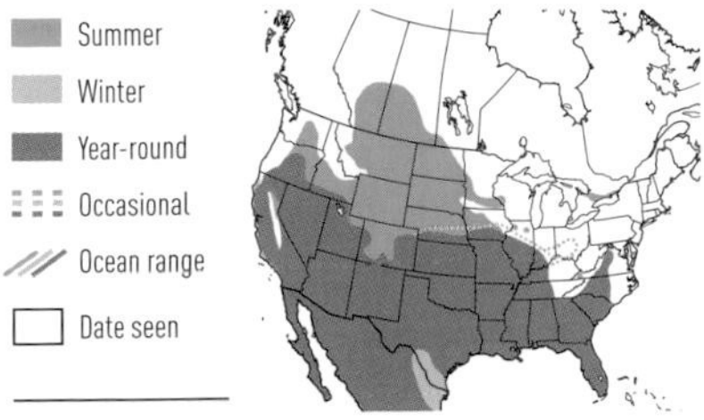

Location ______________________

LOOK FOR: This bold, vocal, and stocky little vireo is our only common songbird with a white iris (hence the name White-eyed). More obvious, however, are the vireo's yellow spectacles around the eyes, its white wing bars, and its pale yellow sides.

LISTEN FOR: The remember-it phrase for the White-eyed Vireo's song is *quick get the beer check!* Or *chik-chik-a-chee-wow!*, given in a harsh, scolding tone, as if the bird is mad about something.

REMEMBER: You are most likely to see a White-eyed Vireo low to the ground in thick cover. Other vireos and warblers are more likely to be found higher up, foraging and singing in trees.

WOW!

Other descriptions of the White-eyed Vireo's song are **Gingerbeer-quick!, Take me to the railroad quick!**, and **Chick of the village!**

▶ *Listen closely to hear the White-eyed Vireo imitate the songs and calls of other birds in its whisper song.*

FIND IT: White-eyed Vireos love old overgrown meadows and viney tangles, but they are not skulkers. You may hear a White-eyed before you see it, and the bird can be coaxed into view with a few simple pishes or squeaks.

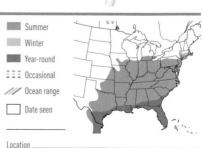

Summer
Winter
Year-round
Occasional
Ocean range
Date seen ______

Location ______

BLUE-HEADED VIREO

Vireo solitarius Length: 5½"

LOOK FOR: It might be a stretch to say this bird has a blue head. But the dark gray head and bright white spectacles (connected eye-rings) and throat are distinctive among our common eastern vireos. It looks chunky headed compared to other vireos.

LISTEN FOR: The Blue-headed Vireo's song is a series of high-pitched, sweet phrases that seem first to ask a question (upward-slurred *see me?*), then answer it (downward-slurred *I'm up here!*). Call is a whiny, inquisitive *chew-wee?*

WOW!

Many Blue-headed Vireos spend the winter in the Southeast, where they survive on berries if insects are not available. Most other vireos spend the winter in the insect-rich tropics.

REMEMBER: Formerly known as the Solitary Vireo, the species was split into three separate species: Blue-headed (mostly eastern), Cassin's (from the Rockies to the Pacific Coast), and Plumbeous (in the inland Southwest).

◀ *Male Blue-headed Vireos do a greater percentage of the work of raising the young than do males of other vireo species.*

FIND IT: The Blue-headed Vireo prefers mixed forest (hardwoods and pines) within its breeding range. Because this species is not as active a forager as some other vireos, it is often first located by sound.

Summer
Winter
Year-round
Occasional
Ocean range
Date seen ______

Location ______

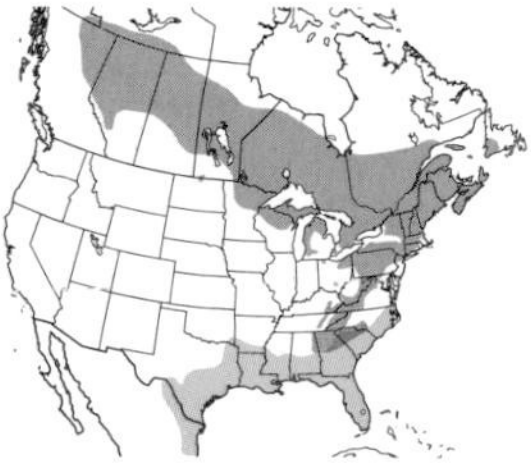

Vireo flavifrons Length: 5½"

LOOK FOR: The Yellow-throated Vireo adds a pair of yellow spectacles to its attractive yellow-throated attire. Look closely at this species, and it appears to have been dipped head-first in yellow paint, leaving the belly white and the back and tail gray.

LISTEN FOR: The song is *three-eight, three-eight, cheerio!*, given in a burry, hoarse voice. Call is a harsh, scolding *chur-chur-chur-chur.*

REMEMBER: The Pine Warbler is similar to the Yellow-throated Vireo, but it has a thinner bill and a lightly streaked belly. The Yellow-throated Vireo's bill is thicker than any warbler's, and its big-headed appearance and slow, methodical movements are very unwarbler-like.

WOW!

Pesticides used to control Dutch elm disease may have caused the disappearance of Yellow-throated Vireos from the tall trees in many towns and city parks. Since this practice stopped, the vireo has experienced population growth.

◀ *The Yellow-throated Vireo often looks as if it is taking an imaginary bath as it hops around, singing its hoarse song.*

FIND IT: Treetop singers and foragers that are more often heard than seen, Yellow-throated Vireos are common in summer in tall hardwood forests, especially in oaks and maples. They winter in the tropics.

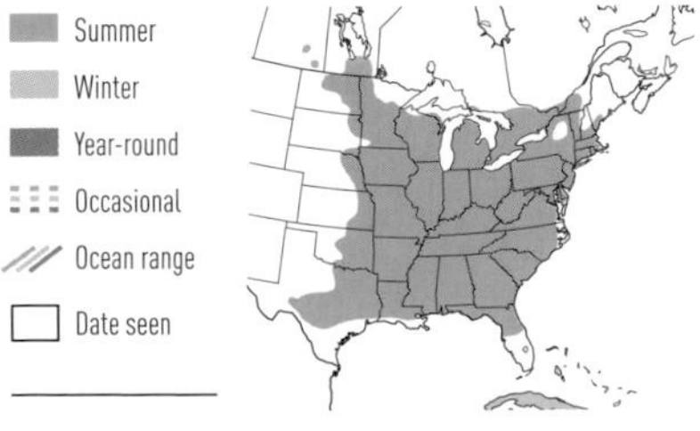

Date seen ______

Location ______

RED-EYED VIREO

Vireo olivaceus Length: 6"

LOOK FOR: The Red-eyed Vireo's best field marks are all on its head. The red eye is not the most reliable one, though it can be seen on most adult birds in good light. Better field marks are the dark gray cap, bold white and gray eye lines, and a longish bill.

LISTEN FOR: Singsong, two- or three-note phrases given once every two seconds or so make up the Red-eyed Vireo's song: *here I am, up here, in this tree, over here.* Call is a harsh, down-slurred *meww!*

REMEMBER: Many of our vireos can appear to be drab look-alikes. The Red-eye's gray cap, contrasting white eye line, plain wings, and long bill help set it apart.

WOW!

The Red-eyed Vireo is the only North American bird with a declarative statement as a Latin name. It means "I am green" in Latin.

◀ *One enthusiastic Red-eyed Vireo was recorded singing more than 27,000 times in one day!*

FIND IT: The Red-eyed Vireo's greenish coloration helps it blend into the foliage, so in summer it is best located by song. In fall, they chase other songbirds and zip around from tree to tree for no apparent reason.

Summer
Winter
Year-round
Occasional
Ocean range
Date seen ______

Location ______

Vireo gilvus Length: 5½"

LOOK FOR: One of our plainest-looking birds, and that's a great clue to its identity. Dull olive-gray above, white below, with a pale eyebrow and no wing bars. Smaller and paler than the Red-eyed Vireo, with a less boldly marked head.

LISTEN FOR: Song is a long unbroken series of rich, husky warbled notes, similar in quality to a Purple Finch's song. Often ends on a higher note. Call is a nasal *quah!*

REMEMBER: This species is best identified by its rich, musical song, which it gives in an unbroken series, unlike the short-phrase songs of our other vireos.

◀ *Warbling Vireo nests are often parasitized by Brown-headed Cowbirds.*

WOW!

The Warbling Vireo's song, as taught to me: **If I see you I will seize you and I will squeeze you till you squirt.** At least it's memorable!

FIND IT: Common in open deciduous woods, mixed coniferous woods, and along wooded waterways in the East. In the West, found in aspen and cottonwood groves, along rivers, and in wooded canyons. Usually heard before it is seen.

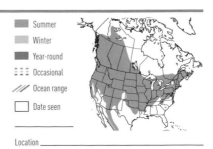

Summer
Winter
Year-round
Occasional
Ocean range
Date seen

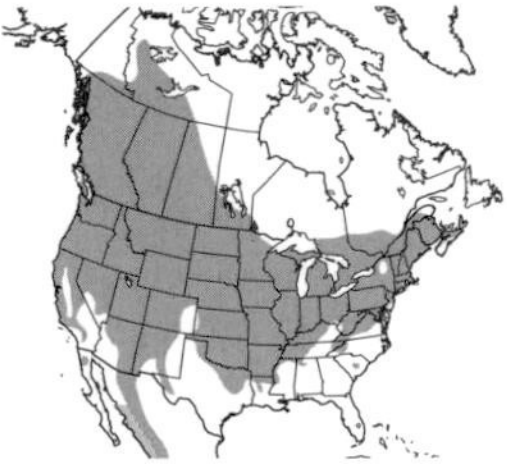

____

Location ______________

WESTERN SCRUB-JAY

Aphelocoma californica Length: 11–11¼"

WOW!
Western Scrub-Jays will sometimes land on deer to remove and eat ticks.

LOOK FOR: A bright blue crestless jay of the West with long all-blue wings and tail, a white throat, dark cheek, blue necklace, and gray-white underparts. Bill is stout and black. Upper back is gray.

LISTEN FOR: Call is a harsh-sounding *kressh-kressh* and a rising *jreee!* Also: *shrek-shrek-shrek-shrek.*

REMEMBER: Western Scrub-Jays living in the interior West are less boldly colored on the face and throat than coastal birds.

▶ *This species readily visits bird feeders for peanuts, suet, fruit, and mealworms.*

FIND IT: Often seen in pairs or small flocks in oak-dominated woods and pinyon-oak woods in a variety of habitats, including canyons, foothills, wooded rivers, and residential neighborhoods.

Summer
Winter
Year-round
Occasional
Ocean range
Date seen
____

Location ____________________

LOOK FOR: Handsome members of the Corvid family (crows and their relatives), Blue Jays are bold and obvious birds much of the year. In flight, the Blue Jay shows a mostly blue back with white inner wingtips (secondaries) and white outer tail tips.

LISTEN FOR: Most common vocalization is a harsh, scolding *jaay, jaay!* Jays also make a variety of other calls, including bell-like whistles and a rusty-gate sound, and they can imitate the calls of raptors, particularly the descending scream of the Red-shouldered Hawk.

REMEMBER: This is the largest blue songbird found in the East. There are several other jays in the West, but only the Steller's Jay is both crested and mostly blue.

WOW!

While a jay may cache thousands of acorns in its lifetime, many are never dug up and consumed but left to grow into oak trees instead.

◀ *Scolding Blue Jays often alert other birds to the presence of hawks or owls.*

FIND IT: Widespread and common all year round in mixed woodlands, city parks, and backyards, where they will come to bird feeders. Northern birds migrate southward in fall, flying in large loose flocks during the day.

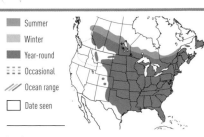

Summer
Winter
Year-round
Occasional
Ocean range
Date seen ____________

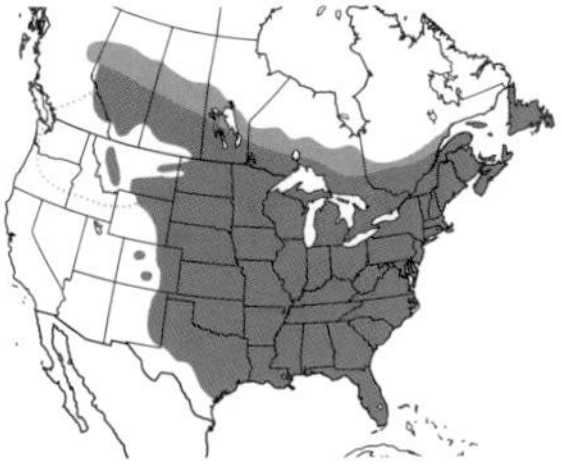

Location ____________

LOOK FOR: A large, dark, crested bird with a large bill and long tail. The front half of the Steller's Jay is black, the back half is dark blue. Flight is direct with long swooping glides.

LISTEN FOR: This very vocal species makes a variety of sounds: a harsh *sharrrr;* a loud, low-pitched *shook-shook-shook;* and a scream that sounds just like that of a Red-shouldered Hawk.

REMEMBER: The all-dark Steller's Jay is unique among its peers. Other jay species of the West lack a crest. The Blue Jay has a pale belly.

WOW!
Steller's Jay could be called Stealing Jay since it often raids the nut caches of Acorn Woodpeckers.

▼ *Steller's Jay flocks often move around by flying in a loose single-file line.*

FIND IT: Common in coniferous forests and mixed pine-oak woodlands of the West, especially in the mountains.

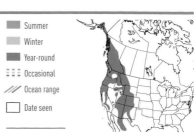

Summer
Winter
Year-round
Occasional
Ocean range
☐ Date seen
____

Location ______________

CLARK'S NUTCRACKER

Nucifraga columbiana Length: 12"

LOOK FOR: Named for its ability to pry pine nuts out of pinecones with its long black bill, the Clark's Nutcracker is mostly tan-gray overall. Its black wings and tail have prominent white patches, most obvious in flight.

LISTEN FOR: A loud, nasal, and harsh *krawww!* and other raspy sounds.

REMEMBER: The similar Gray Jay is more compact-looking with a smaller bill and no white in the wings and tail.

▼ *It's hard to miss the Clark's Nutcracker in flight. Large white patches stand out on the black wings and tail.*

WOW!

A Clark's Nutcracker may bury as many as 30,000 pine seeds in a single fall season. It will retrieve and eat many of them later in the winter.

FIND IT: Prefers conifer forests high in the mountains where it is a year-round resident. Often in small flocks. Perches on treetops and in other prominent places. Like the Gray Jay, will often approach humans for food handouts.

Summer

Winter

Year-round

Occasional

Ocean range

Date seen ______

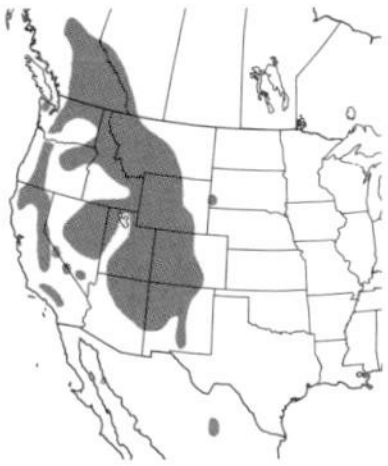

Location ______

GRAY JAY

Perisoreus canadensis Length: 11¼–11½"

LOOK FOR: The Gray Jay is indeed mostly gray with a white forehead and face. But its dark crown and nape and small black bill make it look like a super-sized chickadee. Overall appearance is rounded and "puffed out." Flying birds are plain gray above.

LISTEN FOR: Makes many different sounds, from soft whistles to harsh chuckles and almost gull-like cries. Also a nasal *nyah-nyah-nyah.*

REMEMBER: In the western mountains, the similarly gray Clark's Nutcracker has a large pointed black bill and huge white patches on the tail and wings. Gray Jays are small billed.

WOW!

Among the folk names for the Gray Jay: Whiskey Jack, Camp Robber, Carrion Bird, Grease Bird, and Meat Hawk.

▶ *Gray Jays will boldly follow hikers in the woods, looking for a handout or an opportunity to swipe some food.*

FIND IT: If you are in the woods of the North, it's more likely that a pair of Gray Jays will find *you.* They are year-round residents of spruce and fir forests. Often found near trails, campgrounds, and picnic sites.

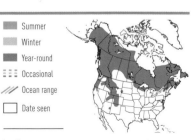

Summer

Winter

Year-round

Occasional

Ocean range

Date seen ______

Location ______

GREEN JAY

Cyanocorax yncas Length: 10½"

LOOK FOR: It's hard to mistake the colorful Green Jay for any other bird. With its lime green back, yellow belly, black throat, and bright violet crown and face, it blends in with its tropical habitat.

LISTEN FOR: A series of harsh, nasal notes: *shack-shack-shack-shack!* Also other weird sounds, like those made by other jay species.

REMEMBER: This is our only green-colored jay.

WOW!

Green Jays help to spread oak trees by burying acorns and never returning for them.

◀ *Many birders get their first look at a Green Jay when one swoops in to eat fruit offered at a feeding station.*

FIND IT: Common in dense scrub woods and mesquite thickets in southernmost Texas, where it is a year-round resident. Usually found in flocks. Visits feeders, backyards, and parks for seeds, fruit, and water.

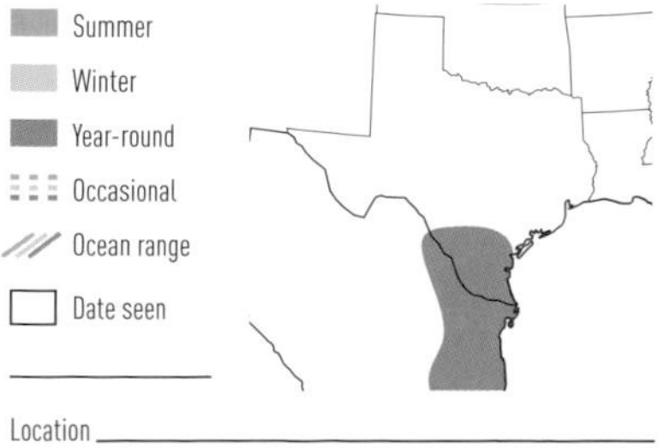

Date seen ______

Location ______

COMMON RAVEN

Corvus corax Length: 24" Wingspan: 53"

LOOK FOR: This large, all-black relative of the American Crow can soar like an eagle, swoop and dive like a falcon, and is adaptable enough to thrive anywhere. In flight it is the wedge-shaped tail that best separates this species from the American Crow (which has a normal fan-shaped tail).

WOW!
Common Ravens have always fascinated humans. Native Americans believe the raven symbolizes death, wisdom, trickery, or evil.

LISTEN FOR: Hoarse, croaking calls (*kraaak, kraak!*), unique to ravens, carry a great distance and are often your first clue to the bird's presence. Also makes a variety of other calls, including rattles, gurgles, and toots.

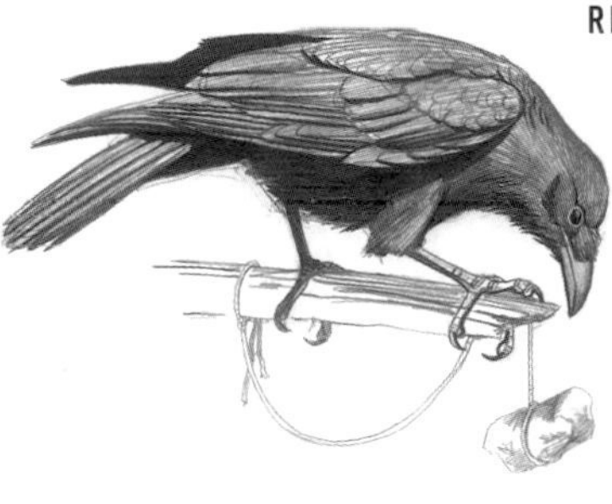

REMEMBER: If you see a Common Raven next to an American Crow, either flying or perched, the raven is the larger of the two. Ravens soar more than crows and look heavier and more powerful in flight.

◀ *Common Ravens are smart enough to solve researchers' puzzles, such as pulling up the string to get the food.*

FIND IT: More common in the North and throughout the West, the Common Raven in the East is found primarily in heavily wooded mountainous areas, above 3,000 feet in elevation.

Summer
Winter
Year-round
Occasional
Ocean range
Date seen

____

Location ______________

BLACK-BILLED MAGPIE

Pica hudsonia Length: 18½–19½"

LOOK FOR: A symbolic resident of the open range in the West, the Black-billed Magpie is named for its least obvious feature. More obvious field marks include its long dark tapered tail, its white belly and shoulders, and the blue-green iridescence on its wings. Flight is smooth and level and clearly shows long tail and white primaries.

LISTEN FOR: A series of harsh calls: *rak-rak-rak-rak* and a rising *maag?*

REMEMBER: Our "other" magpie is the aptly named Yellow-billed Magpie, a resident of the central valleys of California.

WOW!

Until the early twentieth century, Black-billed Magpies were mercilessly hunted down or poisoned as pests and predators.

◀ *Being scavengers, Black-billed Magpies rarely pass up an easy meal of roadkill.*

FIND IT: Common in small flocks in open country with scattered trees or brush, including farmland, rangeland, prairies, and parks. Forages on or near the ground. Often spotted along roadways, where it feeds on carrion (road-killed animals).

Summer
Winter
Year-round
Occasional
Ocean range
Date seen

____

Location ______________

LOOK FOR: Far more common and widespread than the larger Common Raven, the American Crow lacks the raven's large head and bill. In flight, the crow flaps its wings in a smooth, rowing motion and glides, but unlike the raven, it does not soar. A fan-shaped tail further differentiates it (the raven shows a wedge-shaped tail in flight).

LISTEN FOR: *Caa-caa-caa!* is the common call of the American Crow, but this species makes many other sounds, including loud rattles, harsh nasal scolds, and high-pitched, rapid gurgles.

REMEMBER: Another Corvid in the East, the Fish Crow, is often found near water (as the name suggests). Only slightly smaller than the American Crow, Fish Crows are best identified by their call, a nasal, two-noted *ah-ahhh!*

WOW!

American Crows have calls to assemble and disperse the flock, to signal that a predator has been sighted, and to indicate distress, such as when a crow is being attacked by a predator.

▼ *American Crows may travel great distances each day to and from their nighttime winter roosts.*

FIND IT: Incredibly adaptable, the American Crow lives successfully in nearly every habitat, including parks in urban areas. Crows forage on the ground. Outside of the nesting season, crows often gather in huge communal roosts.

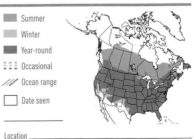

Summer
Winter
Year-round
Occasional
Ocean range
Date seen

____

Location ______________

Tachycineta bicolor Length: 5¾"

LOOK FOR: In bright sunlight, the back of the Tree Swallow glows an iridescent blue-green, contrasting sharply with the pure white throat and belly. The all-white underparts are the best field mark for separating the Tree from other swallows. In late summer young have gray-brown backs and pale gray breast-bands.

LISTEN FOR: The song of the Tree Swallow is surprisingly musical, a series of warbling gurgles: *tia-weet, tia-weet, chur-weet, weet, weet.* Also utters less musical twitters near nest.

REMEMBER: Tree Swallows are among the earliest returning birds in spring. Many spend the winter in the southern U.S., where they survive the cold by switching their diet from insects to berries.

◀ *A male Tree Swallow uses a white feather to court a female.*

WOW!

Tree Swallows line their nests with white feathers for insulation. If you have Tree Swallows nesting near you, toss out a few white craft-store feathers. The swallows will take them to their nests.

FIND IT: In summer, Tree Swallows fly over fields and waterways and nest in tree cavities and birdhouses. Their nesting range is expanding south, helped in part by bluebird nest boxes. In fall, Trees join in huge, mixed-species roosting flocks with other swallows.

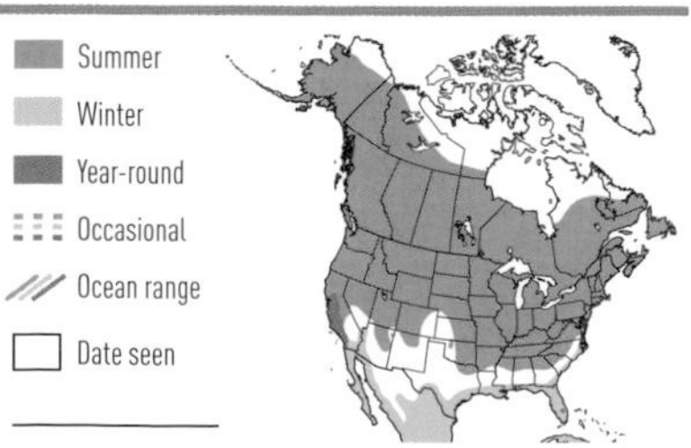

☐ Date seen

____

Location ____________

VIOLET-GREEN SWALLOW

Tachycineta thalassina Length: 5¼"

LOOK FOR: Adult male is bright green and iridescent above and bright white below. The bright white cheeks nearly encircle the eye. In flight the white rump sides show. Female is duller backed, white below. Smaller overall than the Tree Swallow.

LISTEN FOR: Song is a rapid twittering: *chit-chit-chit weet.* Call is a thin *chlip.*

REMEMBER: The Tree Swallow lacks the white face of the Violet-green Swallow.

WOW!

Violet-green Swallows have been observed helping Western Bluebird parents raise their young to fledging. Once they leave, the swallows take over the nest.

▶ *This Violet-green Swallow is nesting in a natural cavity in a tree.*

FIND IT: Common in the West in summer in open woods, mountain forest edges, and in open prairies and towns if nesting sites (boxes, natural tree cavities, rock crevices) are available. Widespread when foraging, often near water. Foraging flocks can be seen high in the air during breeding season.

Summer
Winter
Year-round
Occasional
Ocean range
Date seen

____

Location ______________

LOOK FOR: The Purple Martin is our largest swallow. Adult males are actually a glossy blue, but in direct sunlight their feathers have a purple sheen. Wings and tail are black.

LISTEN FOR: Both male and female Purple Martins sing. Main vocalization is a liquid, warbling chortle given in a series of down-slurred notes: *teer, teeer, teer, jeer-deert, teer!* Males perform a predawn song display in spring, flying high over their nesting colony, hoping to attract females and other martins.

REMEMBER: In flight, Purple Martins resemble European Starlings, but martins fly more smoothly and glide more often. They also show noticeably forked tails.

◀ *Circling high above a chosen nest site, a male Purple Martin attracts potential neighbors with his dawn song.*

WOW!

Hundreds of years ago Native Americans placed hollowed-out gourds on poles for the martins to use. The martins returned the favor by controlling flies and wasps around the village.

FIND IT: Purple Martins are colonial nesters, and in eastern North America, these colonies are made in human-supplied martin houses. In fall, before they migrate to South America, martins form huge roosting flocks at dusk.

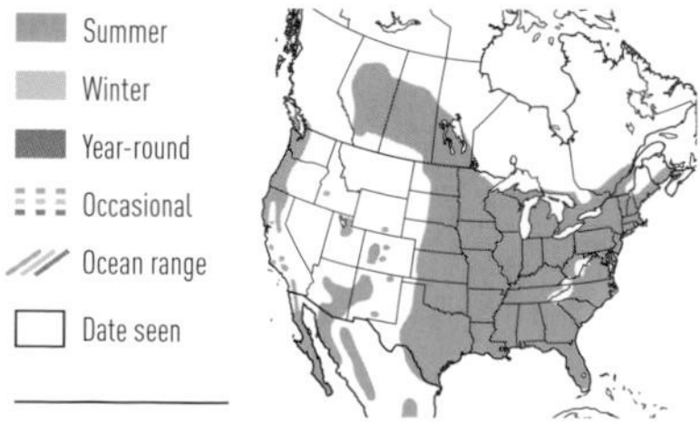

Date seen ____________

Location ____________

Petrochelidon pyrrhonota Length: 5½"

LOOK FOR: The Cliff Swallow looks superficially like a Barn Swallow, but it lacks the Barn's deeply forked tail. The Cliff Swallow has a white forehead, pale collar, and a pale rump that separate it from other swallows.

LISTEN FOR: Cliff Swallows make a series of squeaky noises that sound more like R2D2 from *Star Wars* than a bird. They are most often heard at nesting colonies, where dozens of birds congregate.

REMEMBER: Cliff Swallows have a "headlight" (white forehead patch) and a "taillight" (pale buffy rump), which make it easy to pick the Cliff Swallows from a mixed flock of flying swallows.

Adult

WOW!

A pair of Cliff Swallows makes their nest one tiny mouthful of mud at a time. They may make as many as 2,100 total trips from the mud source to the nest site during the nest's construction.

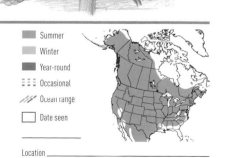

▶ *Cliff Swallows nest in colonies.*

FIND IT: Named for its habit of building its jug-shaped mud nest on cliff faces, today most Cliff Swallows nest under bridges, on dam walls, and inside buildings. This adaptability has expanded the Cliff's range and population.

Summer

Winter

Year-round

Occasional

Ocean range

Date seen

____

Location ______________________________

BARN SWALLOW

Hirundo rustica Length: 7"

LOOK FOR: The graceful Barn Swallow is a common and familiar summer resident across North America. Its deeply forked swallowtail separates this species from our other common swallows. Its rusty face and throat and orange belly help to make this our most colorful swallow.

LISTEN FOR: More vocal than most other swallows, Barn Swallows seem to chatter almost constantly. Main sound is a rapid, chortling series of squeaky notes: *pit-pitpit-pit-pit-pitpit.*

REMEMBER: The Barn Swallow's forked tail is unmistakable. Note that it lacks the white forehead and pale rump of the Cliff Swallow.

WOW!

The Barn Swallow is also found in Europe and Asia, where it is known simply as Swallow. Early European settlers in North America were pleased to find this familiar bird in their new homeland.

◄ *Barn Swallows build their mud nests one billful of mud at a time.*

FIND IT: Barn Swallows can be found wherever there are suitable nest sites, such as old barns and bridge girders. They swoop low over water and follow farm equipment in fields to catch insects startled into flight.

Summer
Winter
Year-round
Occasional
Ocean range
Date seen ______

Location ______

HORNED LARK

Eremophila alpestris Length: 7¼"

LOOK FOR: Horned Larks blend into the ground so well that you may not see a nearby flock until it takes flight, bounding and swooping lightly, showing black outer tail edges. The horns on a Horned Lark may not stick up off the crown. A better field mark is the black facemask and breast band, set off by the yellow throat and pale belly.

LISTEN FOR: Song and calls are high, tinkling notes, like tiny pieces of broken glass blowing in the wind. Their weak songs are often drowned out by the wind.

REMEMBER: Winter Horned Lark flocks may include longspurs, Snow Buntings, and pipits. All these birds have plumages that vary by age and season, so scan these flocks carefully.

▶ *When danger threatens, Horned Lark chicks huddle low in their nests, and a lawn mower might pass right over them without hurting them.*

WOW!

In spring, male Horned Larks hover high above their territories singing their soft, tinkling songs, trying to impress any nearby females.

FIND IT: Summers are spent in the far North on the tundra. In winter, flocks move south, where they prefer open areas such as beaches and farm fields (especially ones with freshly spread manure).

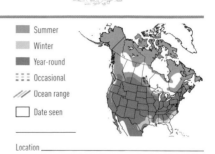

Summer
Winter
Year-round
Occasional
Ocean range
Date seen

Location ____________

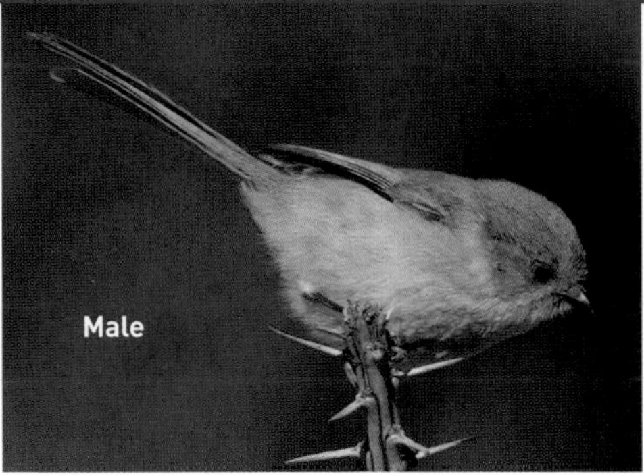

LOOK FOR: A small plain gray bird with a long gray tail, a rounded head and body, and a very short black bill. Female has pale yellow eyes; male has dark eyes.

LISTEN FOR: Calls consistently in a series of high, thin, and scratchy notes: *zrrr, zeeet,* and *tink*. Though it vocalizes constantly, the Bushtit's voice is not loud.

REMEMBER: The Bushtit does have some regional differences in appearance. Birds along the Pacific Coast are browner overall with a brown cap. Some Bushtits in Texas and New Mexico have black "bandito" masks.

WOW!

Bushtits like to party with their friends! Flocks may contain 60 or more Bushtits plus individuals of other species, such as kinglets, warblers, wrens, and chickadees.

▶ *On cold nights, Bushtits may huddle together in a clump to share body heat.*

FIND IT: Found in active flocks (except when breeding) in a variety of wooded habitats, including mixed oak, pinyon, juniper, chaparral, and riparian woods, and parks and wooded residential areas.

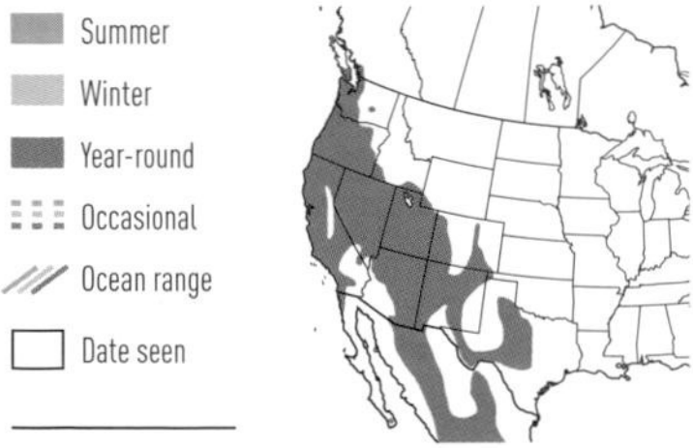

____

Location _______________

VERDIN

Auriparus flaviceps Length: 4½"

WOW!

Verdins know how to survive the desert heat. On hot afternoons they find a shady roost and take a siesta, then resume their activity later in the day when the temps are cooler.

LOOK FOR: A chickadee-sized gray bird with a yellow head and a rusty shoulder patch (which may or may not be visible), the Verdin is an active, vocal bird. Juvenile birds are gray overall.

LISTEN FOR: Calls incessantly: *tsoot, tsoot* or *tweet-tweettweet-tweet!* Calls are variable.

REMEMBER: Of all the small drab gray birds of the southwestern deserts, only the Verdin has a yellow head.

▶ *Male Verdins may build several nests each year. One is used for nesting while others may be used for roosting.*

FIND IT: Though often solitary or in a mated pair, the Verdin's active foraging behavior and constant calling make it easy to notice for a small bird. Found in brushy desert habitats, mesquite thickets, and arid lowlands.

Summer
Winter
Year-round
Occasional
Ocean range
Date seen

____

Location ______________

CAROLINA CHICKADEE

Poecile carolinensis Length: 4¾"

LOOK FOR: The southern cousin of the Black-capped Chickadee is a near look-alike. Carolinas are smaller headed and less buffy on the sides. They also lack the Black-capped's white hockey stick pattern in the wing.

LISTEN FOR: The Carolina Chickadee's song is a high-pitched, sweet *fee-bee, fee-bay* in four or more notes. The call is the familiar *chickadee-dee-dee.*

REMEMBER: Our two common chickadees can be very hard to separate in the field. Range is the best way: in the North it is Black-capped, in the Southeast it's Carolina. Where the ranges meet, it's best to rely on field marks and not voice, since these species learn each other's songs.

WOW!

A Carolina Chickadee nesting in a nest box may hiss at you like a snake and strike at your finger as you check the nest. This behavior is meant to drive off nest predators.

▶ *Carolina Chickadees may alert homeowners when feeders go empty by pecking on windows.*

FIND IT: Though they can be secretive during the breeding season, chickadees are active and vocal, often forming mixed-species feeding flocks with kinglets, nuthatches, and others. They prefer mixed woods, often with large trees.

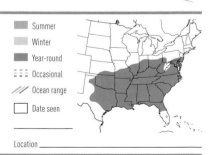

Summer
Winter
Year-round
Occasional
Ocean range
Date seen ______

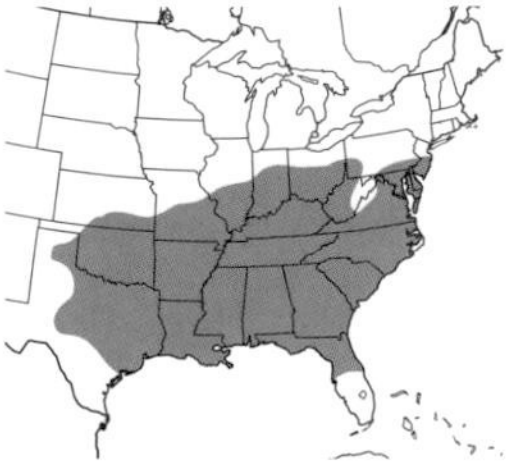

Location ______

BLACK-CAPPED CHICKADEE

Poecile atricapillus Length: 5¼"

LOOK FOR: All of our common chickadees have black caps, but only this one is called Black-capped. Black-caps tend to look bigger headed and chunkier than Carolinas. The white edges of the secondary wing feathers form a hockey stick of white when the wings are folded.

LISTEN FOR: Black-caps have a slower, harsher, and lower-pitched *chickadee-dee-dee* call than the Carolina. Song is usually a two-part *fee-beee!*

REMEMBER: The three Bs of the Black-capped Chickadee: they are Bigger headed, Buffier colored, and have Bolder white edges on their wings.

WOW!

Studies have shown that chickadees actually grow extra brain cells to help them remember where they have cached seeds.

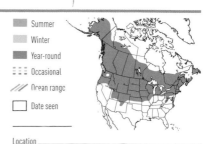

▶ *Black-capped Chickadees attending the nest, one feeding the young, one removing a fecal sac.*

FIND IT: In areas of overlap with the Carolina Chickadee, Black-caps are found at higher elevations. In some winters, Black-caps are found south of their normal range. They prefer mixed woods but are regulars at feeders.

Summer
Winter
Year-round
Occasional
Ocean range
Date seen ____________

Location ____________________________

CHESTNUT-BACKED CHICKADEE

Poecile rufescens Length: 4¾"

LOOK FOR: With its dark reddish back and flanks, this handsome bird looks just like a Black-capped Chickadee that is starting to rust. White cheek patch is narrower in this species than in other chickadees. Chestnut-backed Chickadees along the central California coast have gray (not rusty) sides.

LISTEN FOR: Song is a series of fast, thin chips. Call is a down-slurring *psee-cheer!* Does not have a whistled song like other chickadees.

REMEMBER: In much of the West, if you hear a chickadee's voice, it could be any one of the three widespread species: Black-capped, Mountain, or Chestnut-backed.

▶ *If you live in the range of the Chestnut-backed Chickadee, you might lure them to a feeder with suet.*

WOW!

Nobody knows why this chickadee species got so much more color than our other chickadees did. Just lucky, I guess.

FIND IT: Prefers moist mixed woods, including pines, firs, oaks, and willows. A year-round resident throughout its range.

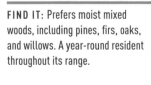

☐ Date seen

____

Location ______________________________

LOOK FOR: Very similar to the more widespread Black-capped Chickadee, but an obvious white line over the eye of the Mountain Chickadee is its key field mark. More subtle clues include the smaller bill and the lack of white feather edges on the wings.

LISTEN FOR: Song is *I see fee-bee!* Call is a hoarse *chick-achickadee-dee-dee.*

REMEMBER: This is the chickadee you are most likely to encounter at higher altitudes in the mountains. Check for the distinctive white eye line.

WOW!

A female Mountain Chickadee will sit tight on her nest. But if really disturbed, she will lunge forward and hiss like a snake to scare away predators.

◀ *When they are not busy breeding or raising young, Mountain Chickadees form mixed-species foraging flocks.*

FIND IT: Common in summer in western mountain forests of conifers and aspen. Often forages in treetops so can be difficult to spot. May move to lower-elevation habitats in winter.

Summer

Winter

Year-round

Occasional

Ocean range

Date seen

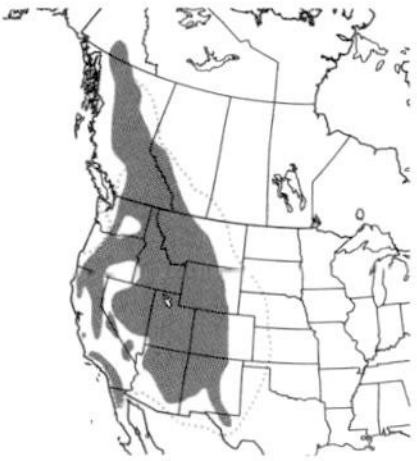

Location ______

LOOK FOR: The Tufted Titmouse's gray-crested head and large black eyes on a pale face give this familiar bird a friendly look. Larger than the chickadees with which it often associates, the Tufted Titmouse shows obvious rusty peach patches on its flanks (sides).

LISTEN FOR: *Peter, peter, peter, peter!* is the clear, whistled song of the Tufted Titmouse. Also utters a harsh chickadee-like scold and a variety of short, sweet, whistled calls.

REMEMBER: There are five titmouse species in North America, but only the Tufted Titmouse is found commonly in the East. A close relative of the Tufted Titmouse called the Black-crested Titmouse is found in central Texas.

WOW!

Tufted Titmice love to line their nests with soft hair. More than one sleeping family pet (and even some humans) have felt a sudden tug as a titmouse boldly steals a bit of hair.

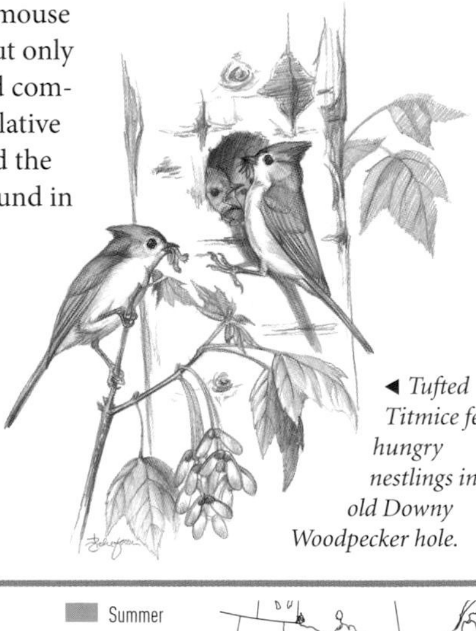

◀ *Tufted Titmice feed hungry nestlings in an old Downy Woodpecker hole.*

FIND IT: Regulars at backyard bird feeders, where they prefer sunflower seeds, peanuts, and suet, Tufted Titmice are both active and vocal. In winter, they may join mixed feeding flocks of chickadees, nuthatches, and others.

Location ____

OAK TITMOUSE and JUNIPER TITMOUSE

Baeolophus inornatus, Baeolophus ridgwayi Length: 5¾"

WOW!

These two species were formerly lumped into a single species called Plain Titmouse. That's still a good name, considering how they look!

LOOK FOR: Both species are short crested and plain gray overall, though Oak Titmouse may be warmer brown. The black eye stands out on the plain face of both species.

LISTEN FOR:

Oak: Song is a loud, whistled *wheetywheety-wheety.* Call is *sissy-sissy-dee.*

Juniper: Song is low pitched and rapid: *jee-jee-jee-jee-jee.* Call: *see-deedeedeedee.*

REMEMBER: The best way to tell these two birds apart is to look at their range maps. It helps to know where you are when you do this, however.

▶ *Like other titmice, the Oak Titmouse holds a nut with its feet while using its bill to crack the shell.*

FIND IT: Oak Titmice are common in oak and oak-pine woodlands. This is the only titmouse species west of the Sierra Nevada. Juniper Titmice are found in oak-juniper and pinyon-juniper woodlands in the Southwest. Both species are year-round residents and usually occur in small flocks.

OAK TITMOUSE

☐ Date seen ____________

Location ____________

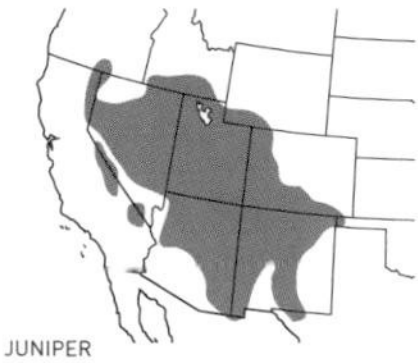

JUNIPER TITMOUSE

☐ Date seen ____________

____

WHITE-BREASTED NUTHATCH

Sitta carolinensis Length: 5¾"

LOOK FOR: The White-breasted Nuthatch has a black crown stripe and a plain white face against which its black eyes stand out. A long, chisel-like bill juts out from its face, and a rusty patch below its tail is obvious. Larger than our other nuthatches, it is also the most commonly seen member of this family.

LISTEN FOR: Song is a nasal series of notes on the same tone: *uhh-uhh-uhh-uhh.* Birders often locate this species by the pounding sound of its bill hacking open a seed against a tree trunk or branch.

REMEMBER: The White-breasted Nuthatch has an undulating flight pattern, much like that of a woodpecker. In flight, it shows contrasting black and white in the tail.

▶ *Squint your eyes and you can see how this nuthatch's threat display resembles an owl's face.*

WOW!

A White-breasted Nuthatch performing its threat display spreads its wings and tail, stands upright with bill pointed up, and waves its body back and forth, looking like a tiny, scary vampire.

FIND IT: The White-breasted Nuthatch prefers mature woods of all kinds. It's a common visitor to bird-feeding stations, where sunflower seeds, peanuts, and suet are its favorite foods.

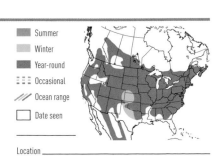

Summer

Winter

Year-round

Occasional

Ocean range

Date seen ______

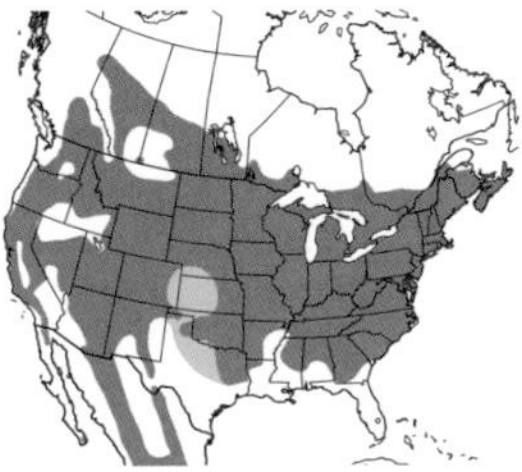

Location ______

Sitta canadensis Length: 4½"

Adult male

LOOK FOR: The orange (not red) breast and black line through the eye set apart the Red-breasted Nuthatch from the larger White-breasted. The male has a black cap and bright orange belly; the female has a blue-gray cap and pale orange belly.

LISTEN FOR: The song of the Red-breasted Nuthatch has been compared to the tooting of a tiny tin horn. The notes are nasal, each one slurring upward in tone: *yenk, yenk, yenk, yenk.*

REMEMBER: Nuthatches have powerful legs and feet and can climb up and down tree trunks and along branches. Red-breasteds look like tiny wind-up toys as they walk along tree trunks and branches. Brown Creepers can only crawl up—never down—because they must use their tails as props to support themselves.

◀ *Red-breasted Nuthatches prefer conifers in all seasons. This one has pried a seed out of a spruce cone.*

WOW!

The name nuthatch describes this family's habit of wedging a seed in a crevice, then hacking the seed open with the chisel-like bill.

FIND IT: Northern coniferous forests and mountains are the summer home of the Red-breasted Nuthatch. In winter, Red-breasteds can be found in a variety of wooded habitats. Like their White-breasted cousins, they will visit backyard bird feeders.

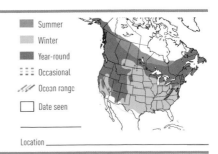

Summer

Winter

Year-round

Occasional

Ocean range

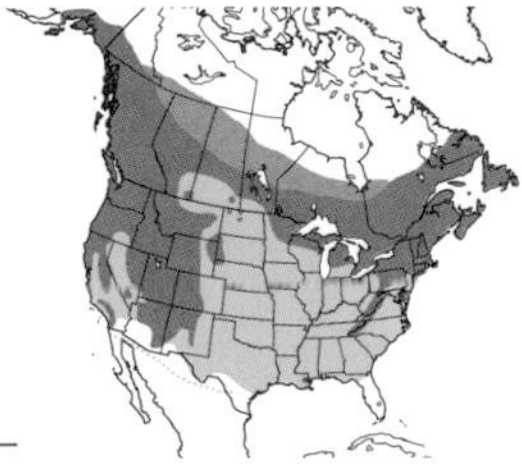

☐ Date seen ______

Location ______

PYGMY NUTHATCH

Sitta pygmaea Length: 4¼"

LOOK FOR: Our smallest nuthatch is aptly named. Its head and bill look big on its small body. Gray back, brown cap, with a black line through the eye. Buffy and gray below. White spot on dark nape.

LISTEN FOR: Often heard before it is seen. Calls are high-pitched, clear squeaks: *pee-deedee or pee-pee!* Calls constantly.

REMEMBER: This tiny treetop-loving nuthatch can be hard to spot. Learning its call will alert you to its presence and help you tell it apart from White- and Red-breasted Nuthatches.

WOW!

In winter flocks of up to 15 Pygmy Nuthatches will roost together in a single tree cavity.

▶ *Pygmy Nuthatches are experts at clambering over pinecones to get at pine seeds.*

FIND IT: Nearly always found in pine forests, where it forages for seeds and insects, often in the highest branches. Often in noisy mixed flocks of chickadees, titmice, and warblers.

Summer
Winter
Year-round
Occasional
Ocean range
Date seen ________

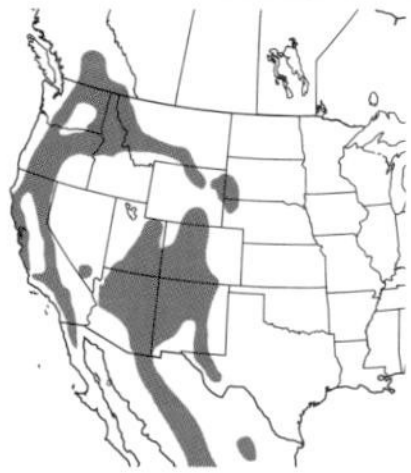

Location ____________________

LOOK FOR: This tiny nuthatch has a brown cap with a white spot on the back of its neck and a bill that looks too large for the bird's size.

LISTEN FOR: If any bird's call ever sounded like a squeeze toy, this one is it. The Brown-headed Nuthatch normally utters a series of high, two-syllable squeaks: *pyee-deet! pyee-deet!* Also gives a single high *queet!* and a burbling series of squeaks.

REMEMBER: A close relative of the Brown-headed is the Pygmy Nuthatch of western pine forests. Fortunately, these look-alike birds are geographically separated, or we'd have a hard time telling them apart.

WOW!

This species has been observed using a tool to get its food! A bird will hold a small piece of bark in its bill and pry up other pieces of bark to get at insects or insect eggs hidden underneath.

▸ *Using a wood chip, a Brown-headed Nuthatch pries up a piece of bark, revealing a juicy spider.*

FIND IT: A year-round resident of southeastern pine forests, the Brown-headed Nuthatch is often found in small family groups and in pairs during the breeding season. In winter they may join mixed feeding flocks.

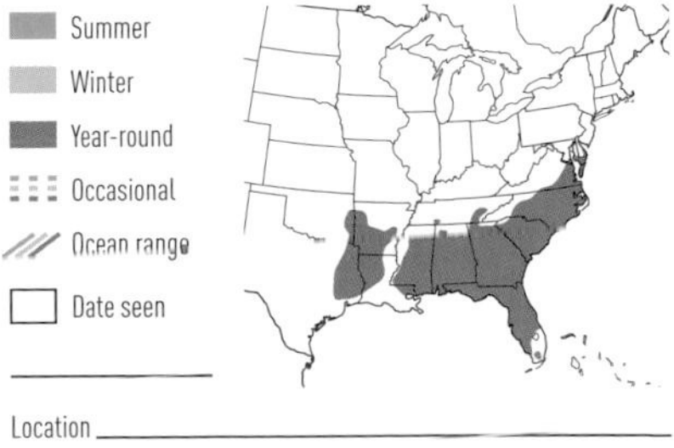

LOOK FOR: Creeping quietly up the trunk of a tree, the Brown Creeper can be extremely hard to spot. Its mottled brown back is the perfect camouflage. The long, stiff tail feathers and thin, curved bill help to separate this species from the nuthatches that forage in a similar fashion.

LISTEN FOR: The creeper's call is a high, thin, single note: *seet!* It is similar to the Golden-crowned Kinglet's call, but the kinglet nearly always gives three *seet* notes. The song is a high-pitched musical one that begins with two *seet* notes and ends with a downward-slurring jumble.

REMEMBER: If you don't hear a creeper's call, you may hear its bill or claws scraping against tree bark before you see the bird.

◀ *Brown Creeper nests are hidden beneath loose tree bark.*

WOW!

Not only do Brown Creepers find their food underneath bark, they nest underneath a large piece of bark still attached to a tree. It's no wonder they look like bark too.

FIND IT: Found in mature woodlands, Brown Creepers creep along the trunks of trees, always climbing upward, very quietly and subtly. When they reach the top of a tree, they fly to the base of another and start over.

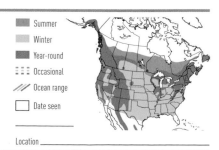

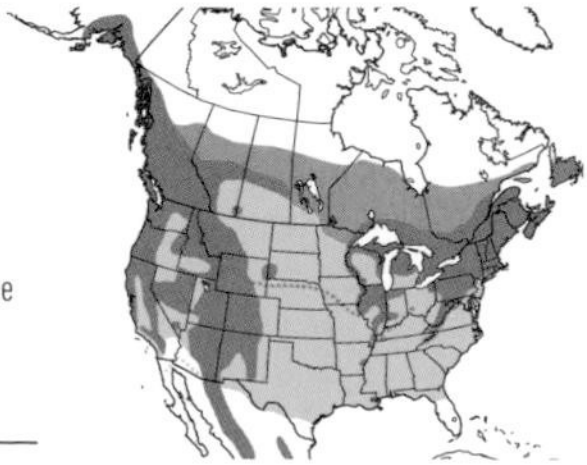

Date seen ______

Location ______

LOOK FOR: This small chocolate brown bird is plain-looking compared to the larger Carolina Wren, but it makes up for this by being very active and very vocal. The House Wren has a faint eye-ring on its otherwise unmarked face.

LISTEN FOR: House Wrens have a song that really warbles. It is a long, high-pitched, liquid burble of notes in a jumble with a few harsh notes at the beginning. Also gives a harsh scold call: *chit-chit-rrr-rrr-rrr!*

REMEMBER: The House Wren vocalizes almost constantly and often cocks its tail upward, much like the Carolina Wren. But the House Wren is smaller, darker brown overall, and lacks the Carolina's white eye line.

WOW!

House Wrens can be very feisty when it comes to nest sites. They will pierce and discard the eggs of bluebirds and others who dare to build a nest in a site the House Wren feels it "owns."

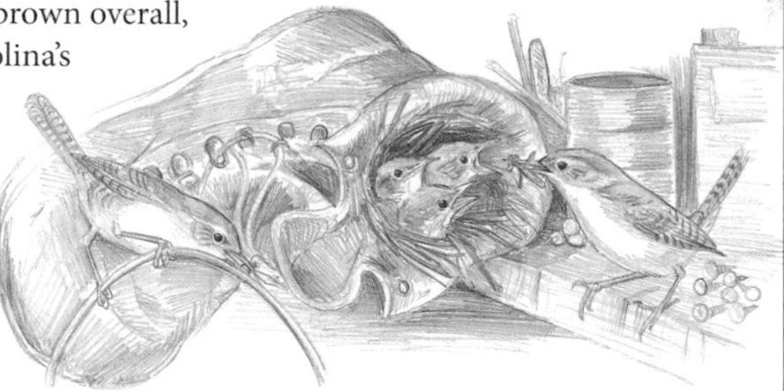

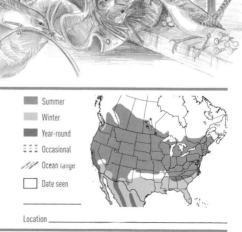

▶ *Anything will do as a nest site for the House Wren—even a discarded boot in a toolshed.*

FIND IT: Most common and vocal in summer, found skulking through thick underbrush and along woodland edges, the House Wren moves (and calls) almost constantly.

Summer
Winter
Year-round
Occasional
Ocean range
Date seen

____

Location ______________

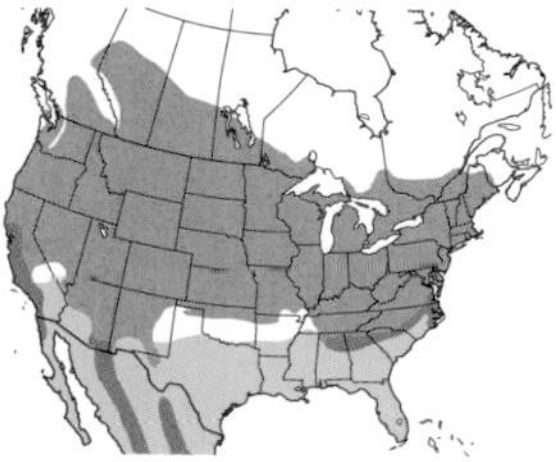

BEWICK'S WREN

Thryomanes bewickii Length: 5¼"

LOOK FOR: A medium-sized wren, plain brown above, pale gray below with a bold white eyebrow line. The long gray-brown tail is rounded at the tip with white corners. While perched, it often fans and switches its tail.

LISTEN FOR: Very vocal. Song is similar to Song Sparrow's: a few high notes, lower trills, rising buzzes, and ending on a trill. Calls include harsh scolding notes, *pit-pit*, and a dry buzz: *dzzzrrrt.*

REMEMBER: The white corners on the Bewick's Wren's tail are diagnostic. The similar Carolina Wren is warmer brown overall with an all-brown tail.

WOW!

The Bewick's Wren is not named for the Buick (though it's pronounced the same). It was named by John James Audubon to honor his friend and fellow artist Thomas Bewick (who lived way before there were any Buicks).

◄ *Bewick's Wrens will take advantage of any handy cavity in which to build their nests, in this case in the wheel well of an old truck.*

FIND IT: Common in the West in scrublands, woodland and riparian thickets, chaparral, and in towns, parks, and backyards with thick underbrush. Declining steeply in eastern parts of range.

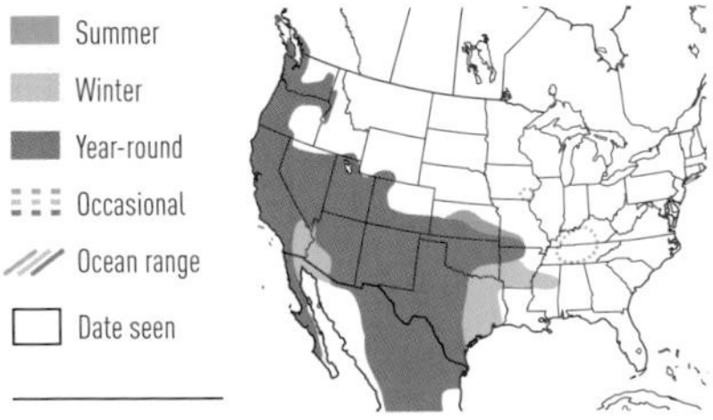

☐ Date seen ____________

Location ____________

CAROLINA WREN

Thryothorus ludovicianus Length: 5½"

LOOK FOR: Our largest eastern wren, the Carolina Wren has a bold white eye line that helps set it apart visually from our other wrens. When perched or foraging, it often holds its tail cocked upward.

LISTEN FOR: Often heard before it is seen, the Carolina Wren's most common call is *teakettle, teakettle, teakettle*. Pairs stay together all year and keep in touch throughout the day by vocalizing back and forth. Carolinas also give a variety of *chink* calls as well as a harsh, rolling metallic scold—*cheerrrrrr-rrr-rrr!*

REMEMBER: The House Wren is an inch smaller than the Carolina Wren, darker brown overall, and lacks the buffy chest and white eye line.

WOW!

Carolina Wrens seem to love nesting near humans. Nests have been found inside garages, in old shoes and empty cans, and even in clothespin bags hanging on well-used clotheslines.

▶ *A Carolina Wren visits its nest inside a toolbox in an old shed.*

FIND IT: Common in brushy thickets and ravines, woodland edges, and backyards, Carolina Wrens readily come to bird feeders, where they eat sunflower seeds and peanut bits, suet, and fruit.

Location ________________

CANYON WREN

Catherpes mexicanus Length: 5¾–6"

LOOK FOR: Rusty brown body and grayish head contrast sharply with white throat and breast. Bill is very long, slender, and slightly down-curved. Small white dots on head, back, and wings are visible at close range. Usually heard before it is seen.

LISTEN FOR: The Canyon Wren is a strong singer, and its song is a series of slurred, clear notes that cascade downward in tone: *tee-tee-tee-tee-tyew-tyew-tyew.* Call is a harsh, burry *jeertt!*

REMEMBER: Our other western wrens lack the contrasting dark belly and white throat and breast of the Canyon Wren.

WOW!

Canyon Wrens use their long thin bills to probe for food in tiny cracks between rocks.

◀ *The Canyon Wren likes to sing from an exposed perch. The song cascades downward, like it's falling from the mountain.*

FIND IT: A year-round resident of steep rocky slopes, cliff faces, canyons, and stone buildings. Often near water. Its loud, ringing song is the best clue to this bird's presence in appropriate habitat.

Summer
Winter
Year-round
Occasional
Ocean range
Date seen

______

Location ______

CACTUS WREN

Campylorhynchus brunneicapillus Length: 8½"

LOOK FOR: Our largest wren, the Cactus Wren is sometimes mistaken for a thrasher. The bold white eyebrow stands out below the dark brown cap. Dark spots below (especially on breast), white streaks above, buffy flanks. Long tail is edged in black and white dashes.

LISTEN FOR: Song is a series of harsh, unmusical notes: *churr-churr-churr-churr,* often speeding up. It sounds like someone trying to start a car. Scold note is *clack!*

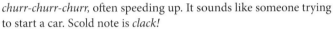

REMEMBER: The somewhat similar thrashers, especially the smallish Sage Thrasher, all have longer tails and plain, unstreaked backs.

WOW!

Cactus Wrens are big, bold, and curious birds. They have been observed picking insects off the front grilles of parked cars.

◄ *The bulky stick nest of a Cactus Wren is an obvious clue to the presence of this species.*

FIND IT: A common year-round resident of arid cactus-mesquite brush-lands in the Southwest. Also found in scrubby sagebrush habitat and coastal scrub. Found in pairs or family groups foraging on or near the ground.

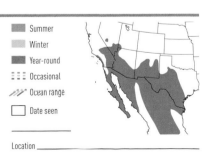

Summer

Winter

Year-round

Occasional

Ocean range

Date seen

Location ______

AMERICAN DIPPER

Cinclus mexicanus Length: 7½"

LOOK FOR: A medium-sized, plump, dark gray bird with a brownish face, short upturned tail, and long pale legs. It will perch momentarily on a midstream rock, bobbing (or "dipping") up and down, before slipping into the water to forage.

LISTEN FOR: Song is surprisingly beautiful, almost mockingbird-like, consisting of a series of rich, clear whistled notes paired with some buzzy tones with phrases repeated. Call is a raspy *zeet!*

WOW!

The American Dipper has built-in swim goggles! It has a third eyelid, called a nictitating membrane. While many animals have this third eyelid, the ones on animals that dive underwater are clear enough to be seen through yet still protect the eyes.

REMEMBER: A good way to spot a dipper is to look for a midstream rock with lots of poop splats on it, and then wait. Dippers have favorite rocks from which to forage and on which to do other things.

▶ *Dippers use their strong legs and short rounded wings to propel themselves underwater in fast-moving mountain streams as they forage for aquatic insects.*

FIND IT: Always found in or near fast-flowing (even roaring) mountain streams in the West. Flight is fast and low over the water. Despite size and constant motion, can be difficult to spot. Resident year-round, but may move to lower elevations in winter.

Summer
Winter
Year-round
Occasional
Ocean range
Date seen

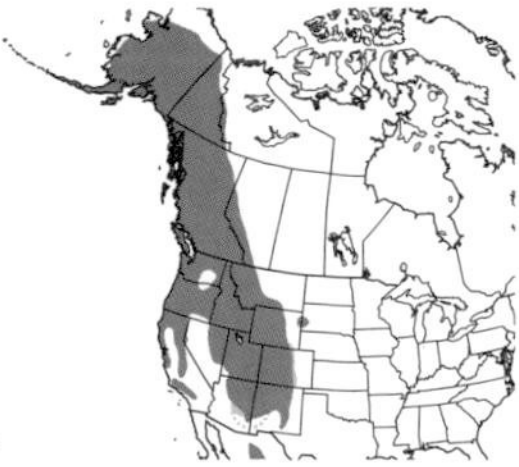

Location ____________

RUBY-CROWNED KINGLET

Regulus calendula Length: 4"

LOOK FOR: The Ruby-crowned Kinglet is slightly larger than its Golden-crowned cousin and lacks the bold markings on the head and face. The Ruby-crowned has an obvious white eye-ring. The male has the namesake ruby crown patch, which can be hard to see because it is raised only when the bird is excited.

LISTEN FOR: The call, *juh-dit-dit!*, is often the first clue to this tiny bird's presence. Song is a long, rich warble.

REMEMBER: No other bird is as tiny, plain, and hyperactive as the Ruby-crowned Kinglet. The eye-ring, plain face and crown, and wing-flitting action set this species apart from the Golden-crowned Kinglet and our warblers and vireos.

WOW!

For such a small bird, the Ruby-crowned Kinglet packs a huge voice. Its song is loud and complex enough to sound like a much larger songbird. Listen for it in the spring.

▶ *Male Ruby-crowned Kinglets in aggressive displays "blow their tops," revealing a burst of red feathers atop their heads.*

FIND IT: The Ruby-crowned Kinglet breeds in the woods of the North, where it sticks to the treetops. In fall and winter, it moves south to wooded habitats, where it forages high and low in the vegetation.

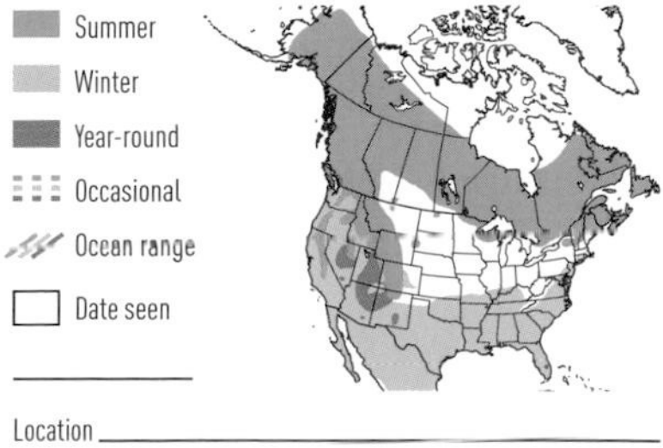

Regulus satrapa Length: 3¾"

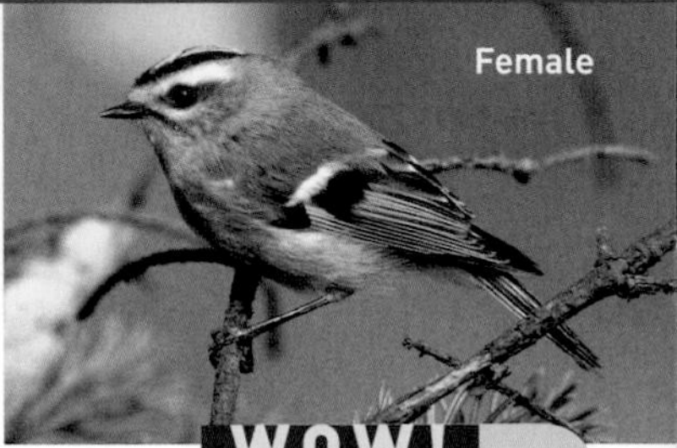

WOW!

Kinglets of both species are able to hover underneath a branch to glean insects off the undersides of the foliage.

LOOK FOR: The tiny Golden-crowned Kinglet flits through evergreen trees, flicking its wings open and shut, hoping to startle an insect into moving. The male has an orange patch on the crown (not always visible), and the female has a yellow one. Its face seems to have white stripes coming out from its tiny black bill.

LISTEN FOR: The call of the Golden-crowned Kinglet is more commonly heard than its song: a high, thin *seet-seet-seet!* The song is a series of thin *seet* notes rising in tone to a thin jumble of notes that falls back down the scale.

REMEMBER: The tiny size, greenish color, and flitting wings can tell you you've got a kinglet. No eye-ring, bold wing bars, and dark crown stripes indicate a Golden-crowned Kinglet.

◀ *Golden-crowned Kinglets can be identified by their habit of hover-gleaning—picking small insects off branches while in flight.*

FIND IT: Nesting in the North or at higher elevations, Golden-crowned Kinglets are most often encountered in winter, when they head south to forage in pines with other kinglets, chickadees, and nuthatches.

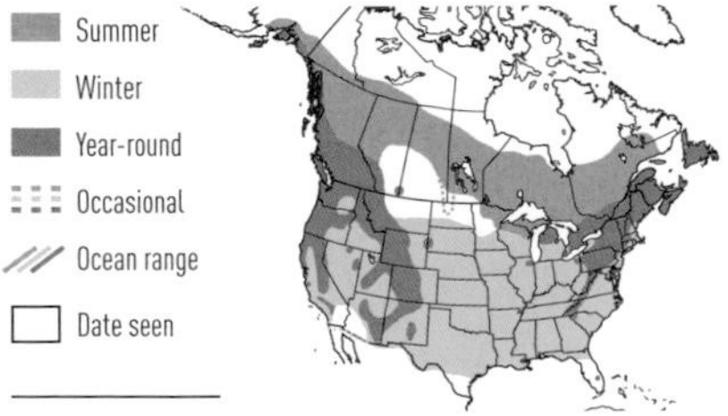

____

Location ______________________________

BLUE-GRAY GNATCATCHER

Polioptila caerulea Length: 4½"

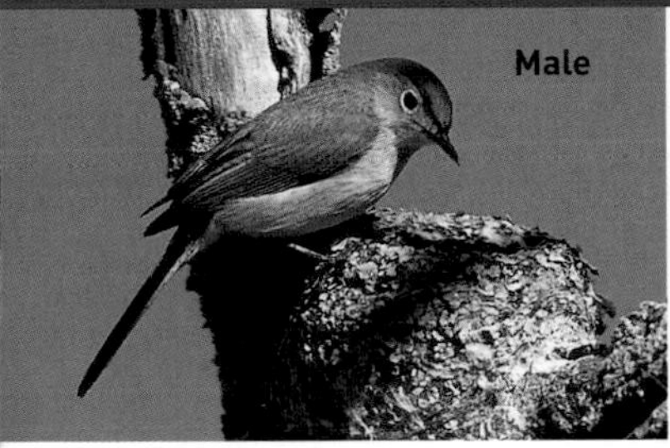

LOOK FOR: This species is well named for both its color and its behavior. Its long tail and white eye-ring on a plain face stand out visually. Its active treetop foraging for gnats and other small insects and its almost constant calling make it easier to spot than many of our small songbirds.

LISTEN FOR: The Blue-gray Gnatcatcher's most common vocalization is its call: a mewing *chee-chee-chee.* Song is a lively and messy jumble of high, nasal notes and short warbles and whistles. Song may occasionally incorporate phrases from other birds.

REMEMBER: The Blue-gray Gnatcatcher is our only gnatcatcher that is common outside of the Southwest.

WOW!

Many consider the arrival of the Blue-gray Gnatcatcher a better sign of spring than the first American Robin. Gnatcatchers appear in many areas as soon as small insects become active.

◀ *Gnatcatcher nests can be easy to spot in spring when the birds are actively building and courting.*

FIND IT: Common in spring and summer in open mixed woodlands and along woodland edges, the Blue-gray Gnatcatcher announces its presence almost constantly with calls, short flycatching flights, and active movements.

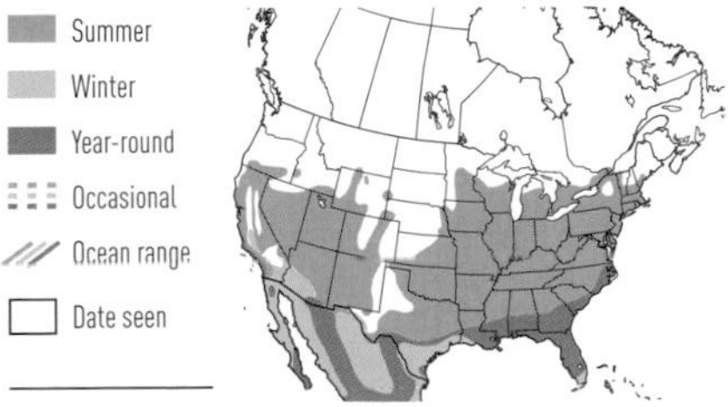

Date seen ____________

Location ____________

Sialia sialis Length: 7"

LOOK FOR: This beautiful member of the thrush family and its relatives in the West (the Western and Mountain Bluebirds) are the only thrushes that nest in cavities. They appear round bodied and round headed when perched. Males are more boldly colored than females. Juvenile birds are blue-gray, spotted with white.

WOW!

Bluebird populations were decimated in the mid-1900s. Bluebird lovers provided nest boxes, helping North America's three bluebird species recover.

LISTEN FOR: The Eastern Bluebird's song is a soft, rich warble given in short phrases: *tur, tur, turley, turley!* Flight call is a two-noted *ju-lee!* When they spot a predator, they utter a down-slurring *tyew!*

REMEMBER: Both Eastern and Western Bluebirds have an orange breast, but on the Western the orange continues onto the back. The Western has a blue throat, while the Eastern's is orange.

► *Eastern Bluebirds prefer to nest in open habitat and will readily use nest boxes, especially where natural cavities are scarce.*

FIND IT: Common in open, grassy settings such as pastures, roadsides, parks, and large backyards, the Eastern Bluebird perches on wires and exposed branches and watches for insect movement below.

Summer

Winter

Year-round

Occasional

Ocean range

☐ Date seen

____

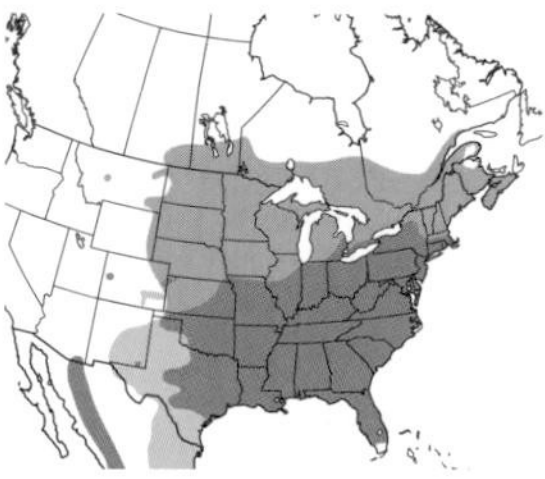

Location ________________________

WESTERN BLUEBIRD

Sialia mexicana Length: 7"

LOOK FOR: The male Western Bluebird is a bold combination of blue and rusty orange. Blue on the throat and belly and rust on the shoulders and back are diagnostic. Female is paler overall with a grayish throat and belly. Juveniles are grayish spotted with white.

LISTEN FOR: Song is simple, down-slurred whistled notes: *tyew. Tyew-tyew!* Call note is harsher chatter: *chack-chack.*

REMEMBER: Similar to the less colorful Eastern Bluebird, but their ranges barely overlap.

WOW!

In winter, when and where juniper berries are abundant, Western Bluebirds may gather in huge flocks to chow down.

▶ *Western Bluebirds will roost communally on cold nights.*

FIND IT: Common in the West in open habitat with scattered trees, including open forests, savanna, mountain meadows, farmland, parks, and golf courses. Often seen in small flocks. Regular user of roadside nest boxes. Winters in habitat with berry-producing trees, such as junipers.

Summer

Winter

Year-round

Occasional

Ocean range

Date seen ____________

Location ____________

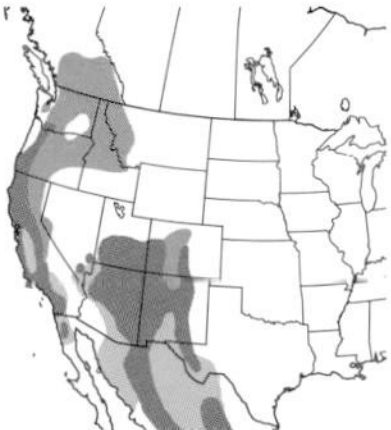

MOUNTAIN BLUEBIRD

Sialia currucoides Length: 7¼–7½"

LOOK FOR: Adult male is pale turquoise blue overall; female is gray below and pale blue on wings and tail. Mountain Bluebirds are more slender and have longer wings and tails than our other two bluebirds.

LISTEN FOR: Song is a series of soft warbles: *too-toodle, too-too-toodle.* Call is a soft *tchack* or *tchack-it.*

REMEMBER: Unlike our other bluebirds, the Mountain Bluebird has no rust or red in its plumage.

WOW!

Because it prefers wide-open spaces, often far from any signs of human activity, the Mountain Bluebird does not get as much competition for nest sites from European Starlings and House Sparrows as its fellow bluebirds get.

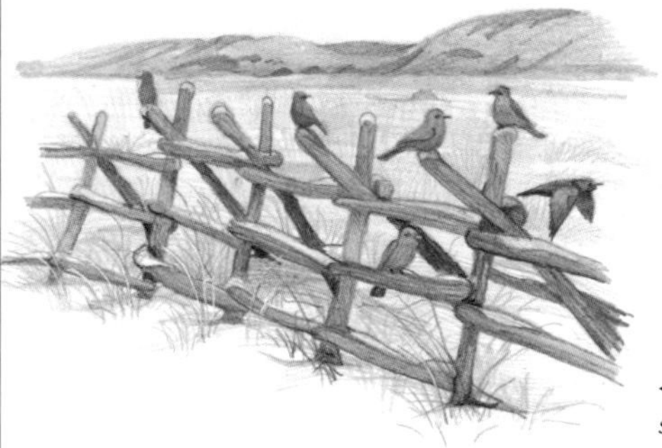

◀ *In winter, Mountain Bluebirds sometimes form huge foraging flocks.*

FIND IT: Prefers wide-open country with scattered trees in all seasons. Often hovers just above the ground looking for insects. Found in mountain meadows and above the tree line, but also in lower habitats, including prairies, sagebrush, and farmland.

Summer
Winter
Year-round
Occasional
Ocean range
Date seen ______

Location ______

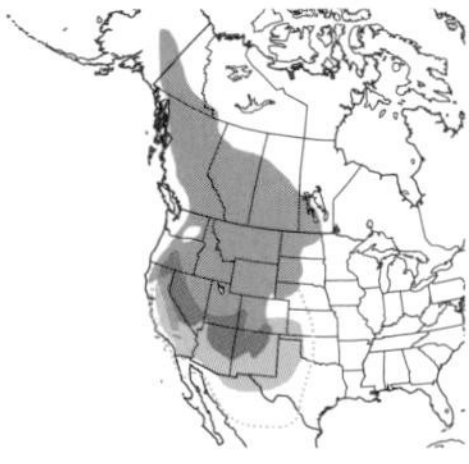

LOOK FOR: A slender gray member of the thrush family, with a prominent white eye-ring and white edges on its long gray tail. Adult Townsend's Solitaires show varying amounts of buff in the wings.

LISTEN FOR: Song is a rich, finchlike warble, slightly hoarse. Call is a high, ringing, single whistled note: *too.*

REMEMBER: The solitaire's short bill, eye-ring, and buff wing patches distinguish it from the Northern Mockingbird and our two shrike species.

▶ *Townsend's Solitaires perch in a very upright posture—a good clue to their identity.*

WOW!

Townsend's Solitaires defend winter territories to protect their food source (berries) from other birds. No wonder they're always alone.

FIND IT: An uncommon and solitary bird of western montane forests, woodland edges, and wooded canyons in summer. In winter, it moves to juniper and pinyon forests, often along streams. Usually seen perched at the very top of a tree or snag, watching for passing insects.

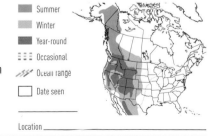

Summer
Winter
Year-round
Occasional
Ocean range
Date seen ____________

Location ____________

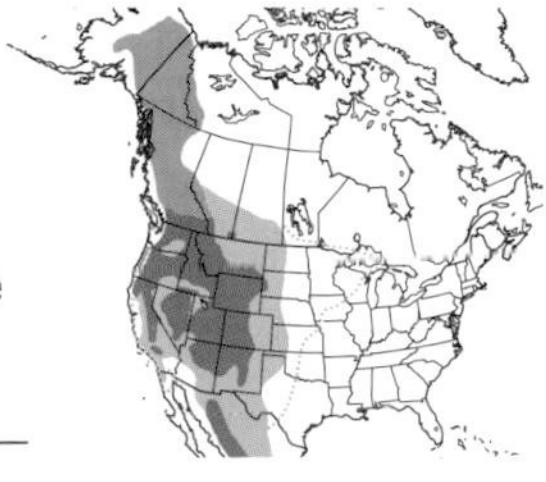

Catharus guttatus Length: 6¾"

LOOK FOR: Several field marks set the Hermit Thrush apart from our other brown woodland thrushes: a lightly spotted breast, a strong buffy eye-ring, a rusty tail, and a habit of raising and lowering its tail when perched.

LISTEN FOR: Many birders consider the Hermit Thrush's ethereal, flutelike song to be among the most beautiful of all bird songs. It starts with a clear, low note and spirals upward in a sweet jumble. Two very different call notes are commonly given: a soft *tchup!* and a nasal, rising *vreee!*

REMEMBER: Each of our six brown woodland thrushes has at least one unique field mark. The Hermit's are its eye-ring, rusty tail, and tail motion.

WOW!

The Hermit Thrush's beautiful song has earned it the folk nicknames American Nightingale and Swamp Angel.

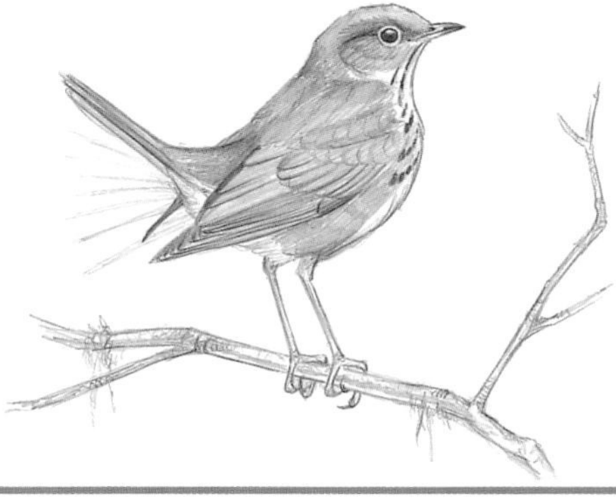

◄ *The Hermit Thrush has a unique habit. It raises its tail suddenly, then lets it fall slowly back to a normal position.*

FIND IT: This is the thrush you are most likely to see here in the winter. They spend the breeding season in northern coniferous forests. In winter and during migration, they can be found in any wooded setting.

Summer
Winter
Year-round
Occasional
Ocean range
Date seen

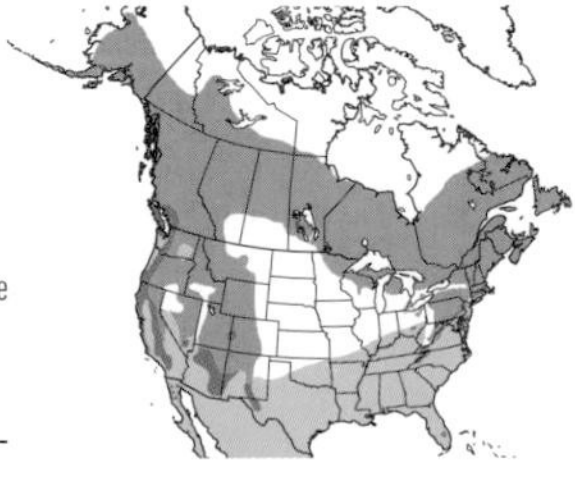

Location

AMERICAN ROBIN

Turdus migratorius Length: 10"

LOOK FOR: The American Robin is so familiar in North America that even nonbirders know its name. Robins spend hours foraging for earthworms in our yards, gardens, and parks. Male robins have darker heads and backs; females are paler overall. Juveniles are spot breasted throughout their first summer.

LISTEN FOR: Song is a rich, slightly hoarse warble: *cheery-o, churlee, cheery-up!* Calls include a loud *see-seet-tut-tut-tut!* and a thin, soft, down-slurred *tseeeet!* as an alarm call.

REMEMBER: In all seasons and at all ages, robins have orange on their breasts, with dark heads and backs.

WOW!

Robins are often considered the first sign of spring, but not all robins leave their home range in winter, so their appearance is not really a sign of spring.

▶ *An adult American Robin jams a juicy earthworm into the gaping mouth of a nestling.*

FIND IT: Robins are found in a variety of habitats, from suburban parks and backyards to mountain meadows, and are most commonly seen on the ground. In winter, they concentrate in the woods where there are berries and other fruits to eat.

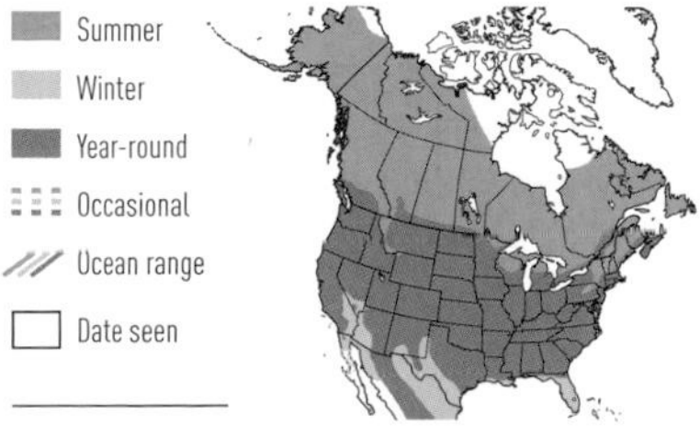

Date seen ____________

Location ____________

LOOK FOR: Looking like an American Robin that put on a bunch of makeup and a black necklace, the adult male Varied Thrush also has orange wing bars and an orange eye stripe. Female is paler overall. In flight it shows an orange central wing stripe.

LISTEN FOR: Song is a long buzzy whistle given in varying tones at short intervals: *tzzzeeee!* Call is a soft *chup.*

REMEMBER: A Varied Thrush may look a bit like an American Robin, but the two species rarely mingle.

WOW!

Every year a few Varied Thrushes get mixed up in fall migration and end up far to the east. This causes eastern bird watchers to go cuckoo.

◀ *Winter flocks of Varied Thrushes eat mainly berries and fruits, but they also forage on lawns for insects.*

FIND IT: Common (but often difficult to see) in dense, wet coniferous forests (spruce, hemlock, fir) of the Northwest where it breeds. In winter, flocks of Varied Thrushes occupy moist thickets, brushy ravines, and coniferous forests. Often heard before it is seen.

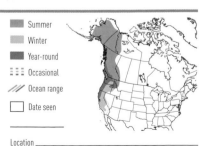

Summer

Winter

Year-round

Occasional

Ocean range

Date seen

____

Location ______________________________

WOOD THRUSH

Hylocichla mustelina   Length: 7¾"

LOOK FOR: The Wood Thrush is the largest and most boldly marked of our brown woodland thrushes. The large spots on the white breast stand out even in deep woodlands. A bright rusty brown back and white eye-ring also stand out.

LISTEN FOR: A beautiful song starting on a clear note or two, followed by a flutey warble: *tu tu ee-oo-laay-yi-yi.* Two common calls are *tu-tu-tu* notes and a staccato *pit-pit-pit-pit* often used as a scold or warning.

REMEMBER: No other brown thrush has the striking breast pattern of the Wood Thrush. It might be more easily confused with the Brown Thrasher, but the thrasher has a long decurved bill, a much longer tail, and two wing bars.

▶ *The Wood Thrush gives its flutelike song in the evening, usually from a well-hidden spot in the mid-canopy.*

WOW!

Forest fragmentation has really hurt the Wood Thrush, allowing nest predators easier access to nests deep in woodlands. As a result, Wood Thrush populations are in decline.

FIND IT: The Wood Thrush breeds in mature deciduous woods, often near water. Easily heard, but can be hard to see in its deep woodland habitat. Look for it foraging on the ground, much like an American Robin.

Summer

Winter

Year-round

Occasional

Ocean range

Date seen ______

Location ______

BROWN THRASHER

Toxostoma rufum Length: 11½"

LOOK FOR: This large, rusty brown bird is often a skulker in thick underbrush. The long decurved bill, twin wing bars, and bright yellow eye help to separate the Brown Thrasher from the Wood Thrush and our other brown woodland thrushes.

LISTEN FOR: Song is a series of warbled and whistled phrases in pairs. The old farmer's rendering of the thrasher's song is *see it see it, pick it up pick it up, dig it dig it, plant it plant it*. Call is a metallic *chaak!* that sounds like two giant marbles cracking together.

WOW!

The spring arrival of the Brown Thrasher on its breeding range is considered by many old-time farmers to signal planting time for certain crops, particularly potatoes.

REMEMBER: The Brown Thrasher is named for its foraging technique. It whips its head back and forth as it walks, using its long decurved bill to thrash leaves and twigs out of the way, hoping to uncover an insect.

▶ *A Brown Thrasher thrashes through the leaf litter to reveal an unlucky earthworm.*

FIND IT: Shrubby woodland edges, brushy old-fields, and hedgerows, even in suburban settings, are the favored haunts of the Brown Thrasher.

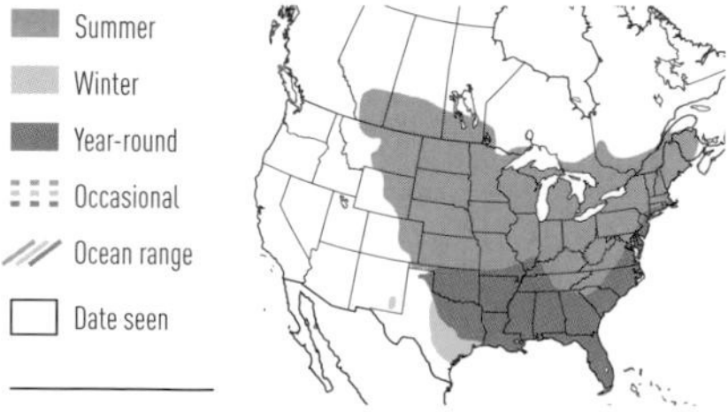

☐ Date seen

____

Location ______________

Dumetella carolinensis Length: 8½"

LOOK FOR: This elegant and slender-looking bird is gray overall with a black cap and a rusty crissum (undertail) patch.

LISTEN FOR: The Gray Catbird has a variable song comprising chortles, squeaks, and short warbles. But it is named for its call note, a mewing (*meeyah!*) that sounds like a cat with a sinus problem. Also utters a loud, staccato *kak-kak-kak-kak!* as an alarm call.

REMEMBER: Except when singing from an exposed perch on the breeding territory, the Gray Catbird and most of our other mimics prefer to remain near the ground and in or near thick cover.

WOW!

The Gray Catbird, as a species, has learned to recognize eggs of the Brown-headed Cowbird, a nest parasite. The catbirds often pierce the cowbird eggs and throw them out of the nest.

◀ *The Gray Catbird is one of our three common mimics—birds that incorporate the songs of other birds into their own songs. The other two are the Northern Mockingbird and the Brown Thrasher.*

FIND IT: Often heard singing or calling before it is spotted in its preferred thick, brushy habitat, the Gray Catbird is a vocal skulker. The mewing call for which it is named is the single best clue to this bird's presence.

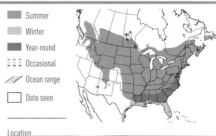

Summer
Winter
Year-round
Occasional
Ocean range
Date seen

____

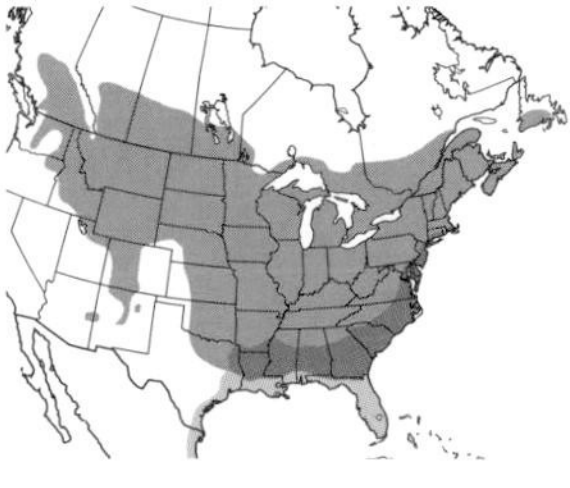

Location ______________

NORTHERN MOCKINGBIRD

Mimus polyglottos Length: 10½"

LOOK FOR: This long, slender, gray, black, and white bird is the king when it comes to variety in its song. It earned its name from its ability to imitate other birds and sounds.

LISTEN FOR: Individual Northern Mockingbirds may use more than 200 different songs and sounds, including those of birds and other animals and mechanical sounds. Most common call is a harsh *tchapp!*

REMEMBER: Few other gray birds flash as much white in flight as the Northern Mockingbird. The Loggerhead Shrike is smaller bodied but stockier in appearance, with a short, thick bill and a black facemask.

WOW!

Unmated young male Northern Mockingbirds will sing all night during the spring and summer, especially when the moon is bright, possibly trying to impress the female mockingbirds.

▶ *Mockingbirds will dive-bomb intruders in their territory, including housecats.*

FIND IT: The mockingbird loves large, open areas, such as lawns and parks surrounded by shrubby undergrowth. As they forage on the ground, mockers fan their wings, flashing the white patches to startle insects into moving.

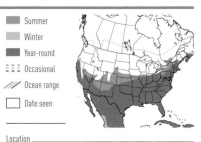

Summer
Winter
Year-round
Occasional
Ocean range
Date seen ____

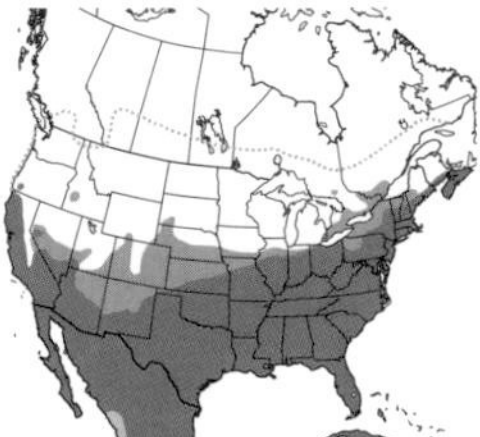

Location ____

CURVE-BILLED THRASHER

Toxostoma curvirostre Length: 11"

LOOK FOR: The long, curved, all-dark bill and the breast with large, dark, round spots help tell the Curve-billed Thrasher from its desert-dwelling thrasher relatives. Has large orange eyes and a long tail. Some birds in Texas have faint white wing bars.

LISTEN FOR: Song is a rich mix of whistles, squeals, squawks, and warbles. Call is a loud, whistled *whit-wheet!* Calls frequently.

REMEMBER: Can be confused with Bendire's Thrasher, which has a shorter bill and triangular (not round) spots on the breast.

◀ *Curve-billed Thrasher pairs remain together all year.*

WOW!

The Curve-billed Thrasher's bill is used to overturn stones and leaves and to dig in the soil in search of food.

FIND IT: The most commonly encountered of our resident desert thrasher species. Found in deserts, arid brushlands, desert canyons, around ranches, and in suburban yards. Usually associated with cholla cactus or prickly pear cactus.

Summer
Winter
Year-round
Occasional
Ocean range
☐ Date seen

____

Location ______________

CEDAR WAXWING

Bombycilla cedrorum Length: 7¼"

LOOK FOR: A warm brown bird, the Cedar Waxwing is named for the small, red, waxy tips on some of its wing feathers. Some waxwings have orange rather than yellow tail tips, a color shift caused by eating certain fruits. Young birds appear streaky and grayer overall.

LISTEN FOR: Waxwings vocalize almost constantly, uttering a high, thin *sreee!* It's not much of a song for such a lovely bird, but it is often your first clue that waxwings are around.

REMEMBER: In the North and in winter, the larger, grayer Bohemian Waxwing may be found with Cedar Waxwings. Bohemians have more white in the wing and an obvious rusty undertail. In flight, Cedar Waxwings may be confused with starlings. The waxwing is sleeker and faster and flocks call constantly.

▼ *Two Cedar Waxwings pass a wild cherry in a bonding ritual used between mates as well as flock members.*

WOW!

Waxwings are sometimes found on the ground, appearing drunk and unable to fly after eating fermented (or too many) berries. They recover their senses after a short period.

FIND IT: Cedar Waxwings are nomads, going wherever the natural berry and fruit crops are plentiful. They occur in flocks at all times except during the late-spring breeding season.

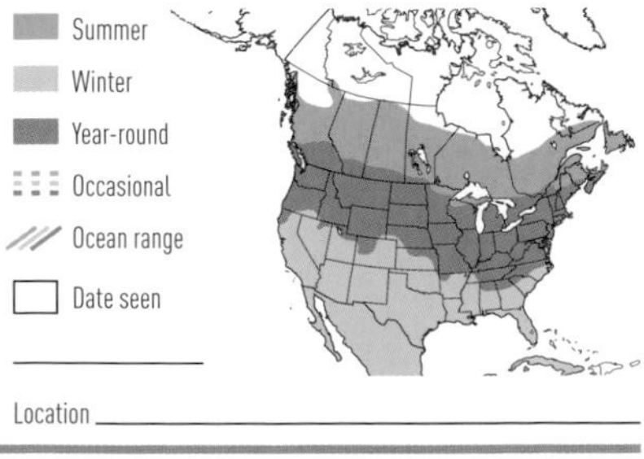

Location ______

PHAINOPEPLA

Phainopepla nitens Length: 7¾"

LOOK FOR: The adult male Phainopepla is a handsome bird: long and sleek and glossy black with a crest and a ruby red eye. In flight, male shows obvious large white wing patches. Adult female is slate gray, also crested and with red eyes. Flight style is flopping and butterfly-like.

LISTEN FOR: Song is a series of odd whistles and harsh phrases sung in a random, disjointed way. Call is an up-slurred, froglike *hoit*.

REMEMBER: No other crested birds are so dark or have red eyes!

WOW!

The name Phainopepla comes from the Greek word meaning "shining robe," in reference to the male's glossy black plumage.

▶ *Mistletoe is the dietary staple of the Phainopepla.*

FIND IT: Common in desert scrub, oak foothills, and mesquite, especially where mistletoe is present. Males especially perch in the tops of trees to sing and flycatch.

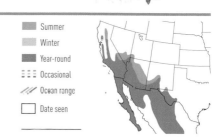

Summer
Winter
Year-round
Occasional
Ocean range
Date seen

Location ______________________

Sturnus vulgaris Length: 8½"

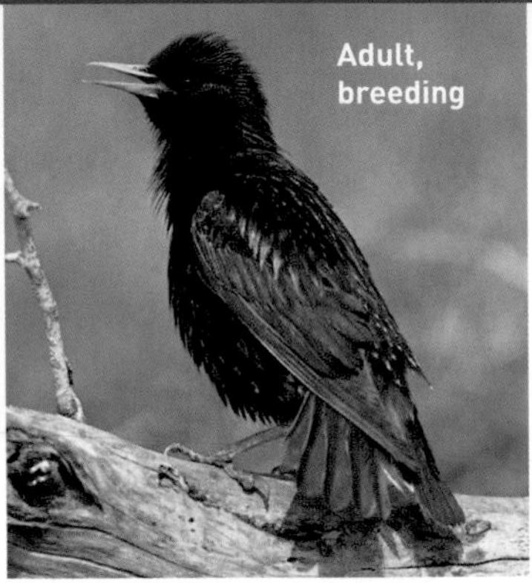

LOOK FOR: Despite its reputation as a stealer of nest cavities and a hog at bird feeders, the European Starling, which looks like a blackbird but is not, is quite lovely in both its glossy breeding plumage and its spangled winter plumage.

WOW!

Starlings love to adorn their nest cavities with shiny or colorful things such as coins, bits of plastic, and other birds' feathers.

LISTEN FOR: Song is a high-pitched jumble of whirs, whistles, and chatter. Excellent mimics, starlings may incorporate other bird songs and even human sounds into their songs. Calls include a metallic *wrrrsh* and *pink!* (often used when a hawk is sighted).

REMEMBER: In flight, the starling can be confused with the Purple Martin: both birds' wings look triangular, but the starling flaps more often and does not glide with the grace of the martin.

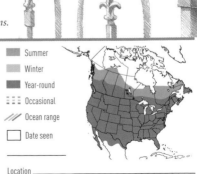

▶ *Starlings prefer to live near humans.*

FIND IT: Introduced from Europe, the European Starling now occurs in almost every North American habitat but has adapted especially well to living around humans. In fall and winter, starlings form large feeding and roosting flocks.

Summer
Winter
Year-round
Occasional
Ocean range
Date seen

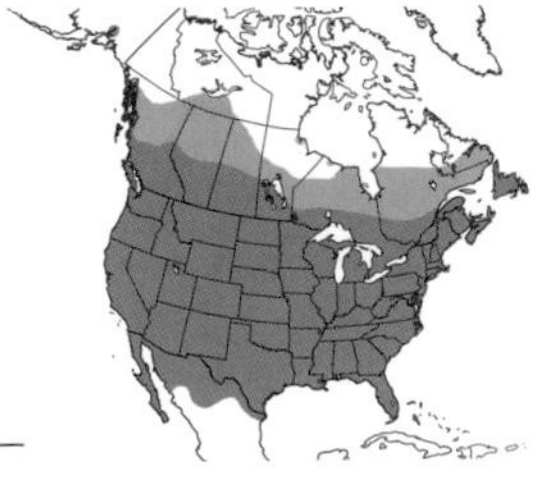

____

Location ______________________________

BLUE-WINGED WARBLER

Vermivora pinus Length: 4¾"

LOOK FOR: Singing from an exposed perch atop a sapling in an old field, the male Blue-winged Warbler is a stunning sight. The male's bright yellow head and body contrast with its blue-gray wings, which have large white wing bars. A slim black line runs through the eye. Females are duller overall. In flight, both sexes flash white outer edges in the tail.

LISTEN FOR: The male Blue-winged Warbler sings *beee-buzzzz!* A frequently given alternative song is a short trill followed by an up-slurred buzz: *ch-ch-ch-ch-tzeee!* Call note is a sweet, sharp *chick!*

REMEMBER: Several warblers have songs similar to the Blue-winged Warbler's. Time spent listening to and learning warbler songs before the birds return in the spring can really pay off.

WOW!

Blue-winged Warblers sometimes interbreed with the closely related Golden-winged Warbler, producing two general hybrids: the Lawrence's Warbler and the Brewster's Warbler.

► *A male Blue-winged Warbler brings his mate a caterpillar while she incubates the pair's eggs.*

FIND IT: Blue-winged Warblers prefer open habitat such as brushy woodlands, woodland edges, and overgrown pastures for nesting and foraging. Territorial males often sing all day long in spring and summer.

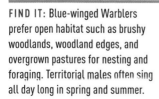

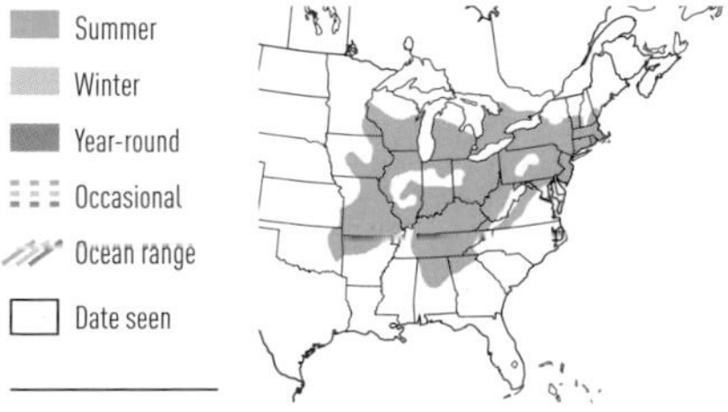

Location ____________________

ORANGE-CROWNED WARBLER

Vermivora celata Length: 5"

LOOK FOR: A small and very plain warbler with a small, finely pointed bill and a light, broken eye-ring. Perhaps the best field mark for this greenish-above, yellowish-below warbler is the yellow undertail.

LISTEN FOR: Song is a fast trill that slows in tempo and drops in tone. Call is a faint *tik!*

REMEMBER: Despite its name, the Orange-crowned Warbler's orange crown is not a regularly seen, reliable field mark.

WOW!

The Orange-crowned is one of our warblers that will visit bird feeders for mealworms, suet, and peanut bits.

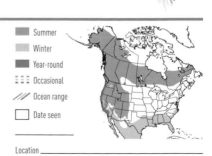

▸ *Orange-crowned Warblers are savvy survivors and will drink sap from sapsucker holes in trees.*

FIND IT: Far more common in the West and North than in the East. In summer, prefers aspen groves, woods with a brushy understory, spruce forest, and streamside thickets. In winter, common in the South in similar habitats as well as in parks and gardens.

Summer
Winter
Year-round
Occasional
Ocean range
Date seen ______

Location ______

WOW!

Northern Parulas love the stringy strands of Spanish moss and other lichens for nest building. The pouchlike nest inside a clump of hanging moss is very hard to spot.

LOOK FOR: A tiny warbler of the treetops, the Northern Parula's (pronounced *PAR-you-lah*) yellow throat and white belly are often all you see as the bird forages high above. But look closely and you may see a yellow lower bill and the reddish black neck band.

LISTEN FOR: The Northern Parula's song is a musical buzz that rises up the scale, ending with a sharp down-slurred note: *zeeeeeee-zup!* One way to remember the pattern of this call is that it climbs up the ladder and drops (or hooks) over the top.

▼ *A male Northern Parula forages in the top of a birch tree.*

REMEMBER: No other widespread warbler has the color combination of the Northern Parula: blue-gray head and back, white eye crescents (like white mascara), and the reddish neck band.

FIND IT: Breeds in large trees often near streams or other bodies of water. A treetop forager, the Northern Parula is best located by its voice, but because it is a small bird, it can be difficult to see.

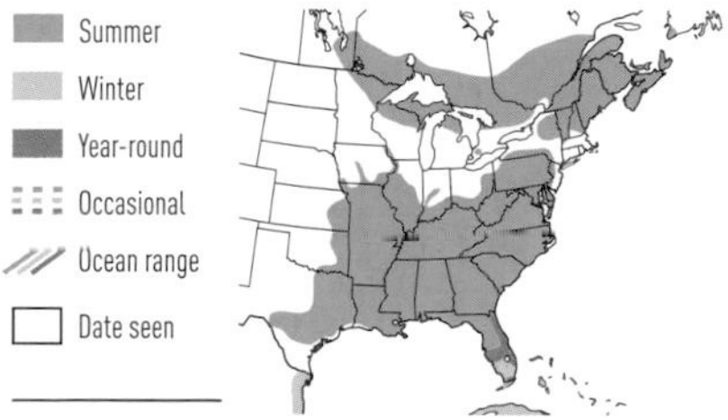

Location ______

Male

LOOK FOR: This bird is well named because nearly every part of it is yellow except for its eyes, bill, and the male's reddish breast streaks. Females lack the reddish breast streaks, and first-year females can appear drab brown.

LISTEN FOR: Song is a variable series of whistled notes: *sweet, sweet, sweet, I'm so very sweet!* Also utters a loud and sharp call: *chip-chip-chip!*

REMEMBER: The Yellow Warbler's dark eyes on a plain yellow face give this species a big-eyed look. The yellow undertail is a good field mark to use when you get only a glimpse of this species. (Most other warblers have white undertails.)

WOW!

Some Yellow Warblers can spot a cowbird egg in their nest. Their response is to start over, often building a new nest on top of the old one.

▲ *A cutaway view of an actual Yellow Warbler nest, including eggs laid by a Brown-headed Cowbird.*

FIND IT: Yellow Warblers love willow trees, especially near water. But they can be found foraging in a variety of trees, usually at midlevel, making them one of our easiest-to-see warblers. Most birds winter in the tropics.

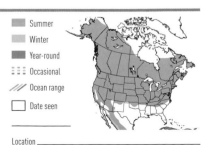

Summer
Winter
Year-round
Occasional
Ocean range
Date seen ______

Location ______

MAGNOLIA WARBLER

Dendroica magnolia Length: 5"

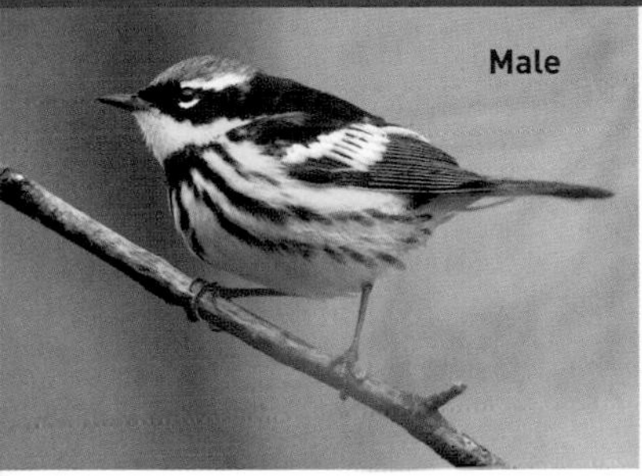

WOW!

The Magnolia Warbler was named by early American ornithologist Alexander Wilson, who first saw this species flitting through a magnolia tree during spring migration in 1810.

LOOK FOR: The male Magnolia Warbler has one of almost every field mark common to our songbirds: eye line, mask, necklace, streaky sides, wing panels, rump patch, tail spots, and black band across tail tip, making it one of our easiest warblers to identify.

LISTEN FOR: Magnolia Warblers sing a sweet but weak-sounding *pretty-pretty Maggie!* Call note is a very sweet *chet!*

REMEMBER: With the Magnolia Warbler, it's all about the white tail spots—and they seem to love to show them off. Females and fall males lack the breeding male's bold facial markings, but all Maggies have the lemon breast, bold black streaks on flanks, and clean white "underpants" on undertail coverts.

▼ *A territorial male Magnolia Warbler flits high in the spruces, flaring its white-paneled tail in a signal to rivals and potential mates.*

FIND IT: Maggies breed in young coniferous forests. They are commonly seen as migrants in almost any habitat in the East (but only rarely in magnolias). They often forage in low, shrubby vegetation during migration and are active and easy to spot.

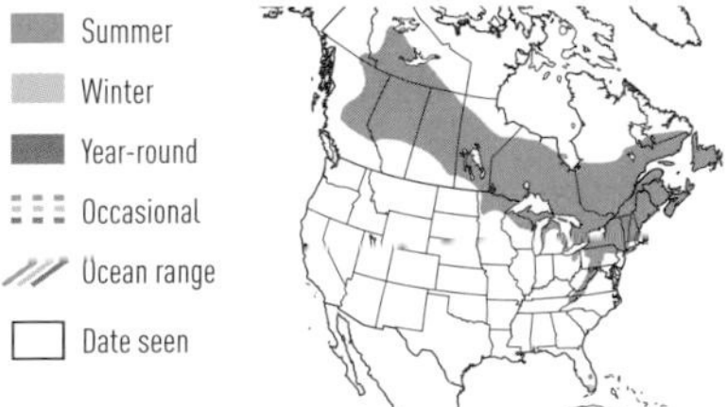

Location ____________

YELLOW-RUMPED WARBLER

Dendroica coronata Length: 5½"

LOOK FOR: The Yellow-rumped Warbler is well named for its most obvious field mark. Older field guides (and many birders) refer to this species as the Myrtle Warbler (eastern form) and the Audubon's Warbler (western form). Both forms have a yellow rump and white tail patches, but Audubon's has a yellow throat and Myrtle has a white throat.

WOW!

Yellow-rumped Warblers are hardy birds, adapted to live on berries (especially wax myrtle fruits) when cold weather makes insects scarce.

LISTEN FOR: Song is a weak trill, usually trailing off toward the end and dropping in tone: *tee-tee-tee-brr-brrbrr!* The call note, a loud *tchep!*, is often a better year-round clue to the presence of this species.

REMEMBER: Cape May and Magnolia Warblers also have yellowish rumps, but they're not as obvious as the butter butt on the Yellow-rumped Warbler.

▶ *A Yellow-rumped Warbler eating wax myrtle fruit in winter.*

FIND IT: Yellow-rumped Warblers spend summers in the northern coniferous forests and at higher elevations. They are very active birds, flitting from tree to tree, flashing their yellow rump patches as they move.

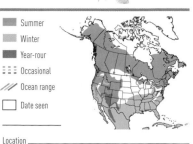

Summer
Winter
Year-rour
Occasional
Ocean range
Date seen

Location ______________________

BLACK-THROATED GRAY WARBLER

Dendroica nigrescens Length: 5"

LOOK FOR: Bold black-and-white pattern of the adult male's head and the gray back suggest a chickadee. Has heavily streaked chest and flanks and white wing bars. Between the eye and bill is a small spot of yellow. Adult female has a white throat.

LISTEN FOR: Song is buzzy and rising: *zeedle-zeedle-zeedle-zeezee* or *Don't ya think I look so pretty?* Call is a strong *tchup.*

REMEMBER: This is the only western warbler that appears to be all black and white.

WOW!

This western warbler occasionally shows up in the East, heating up the rare-bird alerts.

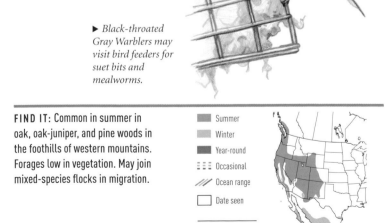

▶ *Black-throated Gray Warblers may visit bird feeders for suet bits and mealworms.*

FIND IT: Common in summer in oak, oak-juniper, and pine woods in the foothills of western mountains. Forages low in vegetation. May join mixed-species flocks in migration.

Summer
Winter
Year-round
Occasional
Ocean range
Date seen

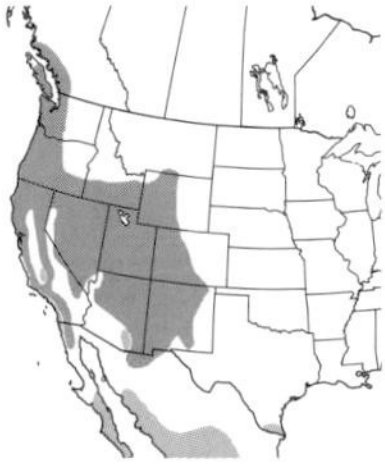

____

Location ______________

TOWNSEND'S WARBLER

Dendroica townsendi Length: 5"

LOOK FOR: Adult male has a boldly marked black and yellow head, with a black cheek and throat. Female has a yellow throat and gray, not black, cheeks. Both sexes have a yellow chest with heavy dark streaking and a white belly.

LISTEN FOR: Song is variable, but usually contains a series of high, thin whistles, ending in buzzier notes: *sleazy-sleazy-cheezy!*

REMEMBER: A similar western warbler, the Hermit Warbler, has a plain yellow face and crown.

◀ *Overwintering Townsend's Warblers often join mixed-species feeding flocks, which may include chickadees, nuthatches, and others.*

WOW!

Birds have evolved foraging strategies that allow them to coexist with their neighbors. The Townsend's Warbler's foraging strategy focuses on the food at the very tops of coniferous trees.

FIND IT: Prefers conifers or mixed conifer-deciduous woods in summer; especially fond of tall trees. In migration and winter (some birds do not migrate) may be found in other habitats.

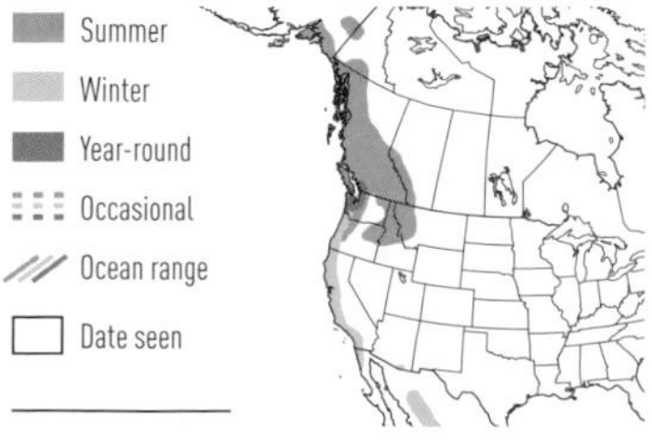

Location ______

BLACK-THROATED GREEN WARBLER

Dendroica virens Length: 5"

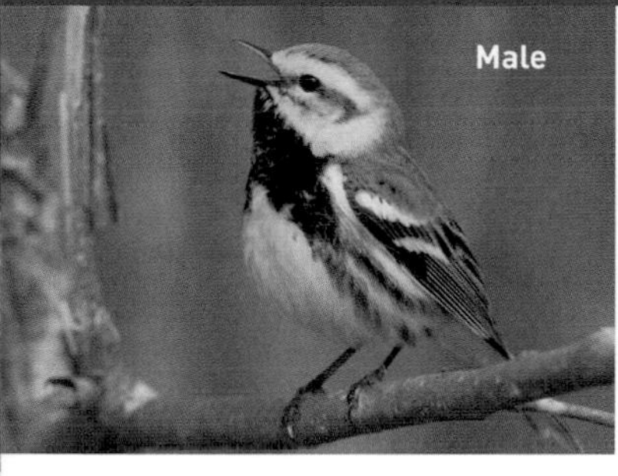

LOOK FOR: It's the golden face patch and black throat that are the most noticeable features on the male Black-throated Green Warbler. Only after seeing these field marks do you notice the green upper back, white wing bars, and olive lines around the eyes. Females lack the black throat. In flight, white outer tail feathers are obvious.

LISTEN FOR: Two typical versions of the Black-throated Green Warbler's song are *zee-zee-zee-zee-zee-zoo-zeee!* and *zoo-zeee-zoozoo-zeee!* Both are buzzy and both jump between two distinct notes. Call note is a thin, sharp *tchet!*

REMEMBER: Many birders refer to this species by the nickname BTG. This is far easier to say than the bird's entire name.

WOW!

One unique phrase used by some birders to remember the BTG's song is **Trees, trees, murmuring trees.**

▶ *A Black-throated Green Warbler searches for insects on leaves in a foraging technique known as gleaning.*

FIND IT: Black-throated Green Warblers nest in coniferous forests and at higher elevations in eastern mountains. This species is a common spring and fall migrant throughout the East, occurring in any wooded habitat.

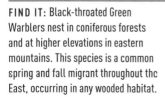

Summer

Winter

Year-round

Occasional

Ocean range

Date seen ______

Location ______

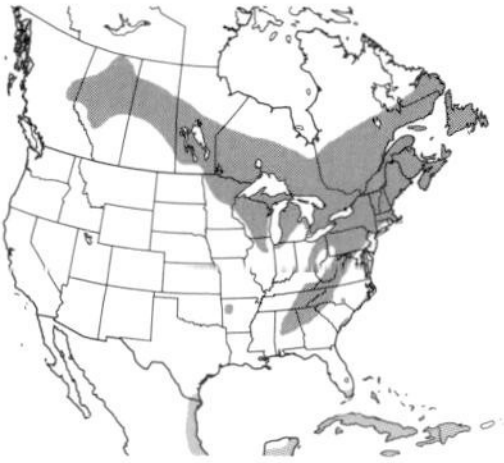

PRAIRIE WARBLER

Dendroica discolor Length: 5½"

WOW!

Prairie Warblers may nest in loose colonies of several breeding pairs—very unusual for warblers. Some lucky males in these colonies may have multiple female mates.

LOOK FOR: The Prairie Warbler is a small warbler with a lemon yellow breast and bold black side streaks. When foraging, Prairie Warblers pump and flit their tails, and in flight their tails show obvious white outer edges. Females wear a duller version of the male's plumage.

LISTEN FOR: The Prairie Warbler's song is a distinctive series of buzzy notes rising up the scale: *zree-zree-zee-zee-zee-zee!* The song can be variable but usually speeds up toward the end. Call note is an emphatic *chek!*

REMEMBER: The dark semicircle below the Prairie Warbler's eyes (most obvious on males) is a distinctive field mark. The tail twitching is another good field mark for this species.

▶ *Male Prairie Warblers often sing their rising song from the top of a short tree.*

FIND IT: Not typically found on the prairie, Prairie Warblers prefer young woods, shrubby woodland edges, and low thickets. They often forage low, flicking their tails and flashing their white tail edges.

Summer

Winter

Year-round

Occasional

Ocean range

☐ Date seen

____

Location ______________________________

LOOK FOR: The Palm Warbler's rusty cap and pale (often yellow) eyebrow in breeding plumage are great field marks for this species. An even better field mark is the Palm's constant tail pumping. In all seasons, as a Palm Warbler pumps its tail, yellow undertail feathers are visible, even on the dullest fall and winter birds.

WOW!

Two folk names for the Palm Warbler, which refer to its behavior, are more accurate names: Wagtail Warbler and Tip-up Warbler.

LISTEN FOR: Palm Warblers sing a weak trill: *tre-tre-tre-tre-tre-tre!* Call note is a crisp *tchit!* that sounds more metallic and sharper than a Yellow-rumped Warbler's chip.

REMEMBER: Palms can range in color from very yellow to very pale, so it's best to rely on the tail-pumping behavior and yellow undertail to identify this species.

▶ *The Palm Warbler's constantly wagging tail, yellow undertail coverts, and habit of walking on the ground are distinctive.*

FIND IT: Rarely associated with palm trees, the Palm Warbler nests in far northern bogs. Palms are a more common sight during migration, when they forage in low vegetation and on open ground.

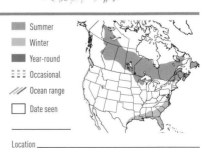

Summer
Winter
Year-round
Occasional
Ocean range
Date seen

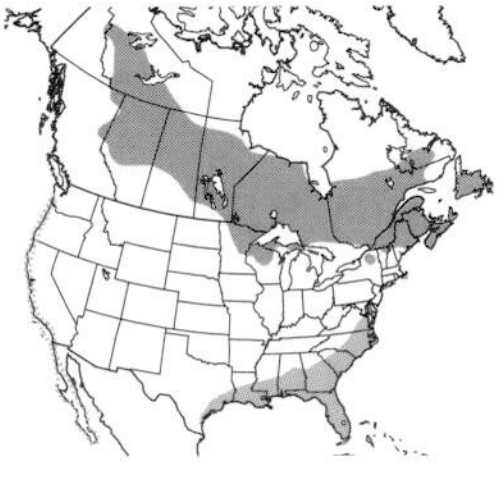

____

Location ______________

BLACK-AND-WHITE WARBLER

Mniotilta varia Length: 5¼"

LOOK FOR: Crawling like a zebra-striped nuthatch along a tree branch or trunk, the Black-and-white Warbler makes its living gleaning insects from bark. It does not change its streaky black-and-white plumage noticeably between seasons. Its bill is long and decurved compared to other warblers.

LISTEN FOR: The song sounds like a squeaky wheel turning: *weecy-weecy-weecy-weecy.* Call note is a sharp, thick *chik!*

WOW!

Two other names for this species are Pied Warbler (for its black-and-white appearance) and Black-and-white Creeper (for its foraging behavior).

REMEMBER: The Blackpoll Warbler and the Black-throated Gray Warbler look similar to the Black-and-white. In spring and summer, the Blackpoll male has a bold black cap and more subtle streaking than the Black-and-white. The Black-throated Gray Warbler has a bold facial pattern and a clear white belly.

◀ *Extra long and strong hind toes allow the Black-and-white Warbler to hang head-down like a nuthatch.*

FIND IT: Creeping along branches and trunks of large trees in old mixed woodlands, the Black-and-white Warbler can be less obvious than many other warblers. Its regular singing and constant motion are helpful in locating this species.

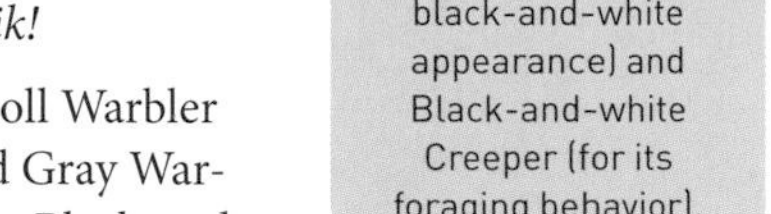

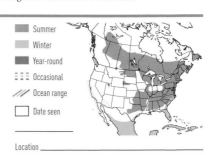

Location ______________________

AMERICAN REDSTART

Setophaga ruticilla Length: 5¼"

WOW!

In Latin America, where the American Redstart spends its winters, this species is called Candelita, the little torch.

LOOK FOR: It's hard to mistake the male American Redstart for anything else. Females and young (first-year) males are gray-green overall with bright yellow patches at the shoulders, wings, and tailbases. Like many other warblers, the redstart is active when foraging, flitting from branch to branch, often opening its wings and tail.

LISTEN FOR: American Redstarts have a highly variable song. Sometimes it's *see-see-see-me-up here!* Other times it's *weeta-weeta-weeta-weet-zeet!* Often they sing one song and then the other. Call note is a clear, sweet *tchip!*

REMEMBER: You might almost confuse the male American Redstart with a miniature version of an Orchard or Baltimore Oriole, but those birds are much larger than the redstart.

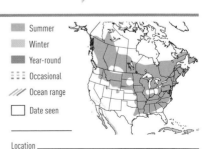

▲ *An American Redstart vaults into the air after a leafhopper.*

FIND IT: Redstarts prefer young woods and scrubby woodland edges for breeding and foraging. They are active singers and foragers, often zipping out from the treetops to catch a flying insect.

Summer
Winter
Year-round
Occasional
Ocean range
Date seen

____

Location ______________

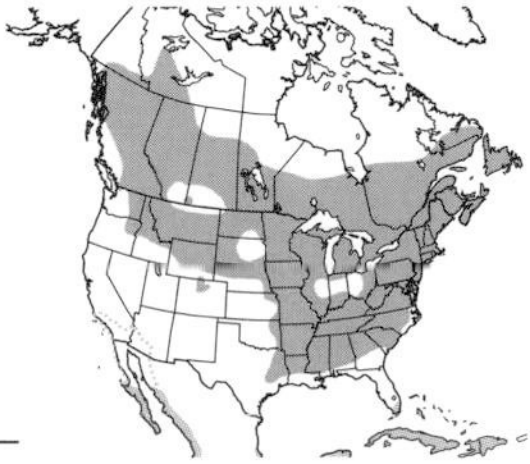

LOOK FOR: This large ground-loving warbler is named for the domed nest (shaped like an outdoor oven) it builds on the forest floor. Males and females look alike and retain their coloration all year.

LISTEN FOR: Among our most distinctive warbler songs, the Ovenbird's song is *tea-CHUR, tea-CHUR, tea-CHUR, tea-CHUR,* given in a loud, ringing voice that starts softly and gets louder at the end. Call note is a loud *chep!*

REMEMBER: Four other large brownish warblers share the Ovenbird's general habit of living near the ground. The Northern and Louisiana Waterthrushes, Worm-eating Warbler, and Swainson's Warbler all have eye lines rather than the Ovenbird's white eye-ring.

◀ *The Ovenbird's domed nest is often constructed under a preexisting tent of vegetation, such as Christmas fern.*

WOW!

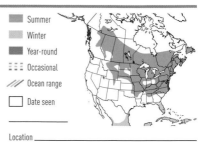

Territorial male Ovenbirds sing a nighttime flight song during spring and summer. He flies above the treetops, singing loudly, then drops back into the dark woods.

FIND IT: Although its song is loud and obvious, this earth-colored bird can be hard to locate. Usually seen walking or foraging with its tail pointed up. Singing males may be perched near the trunk of a tree, just below the canopy.

Summer
Winter
Year-round
Occasional
Ocean range
☐ Date seen

____

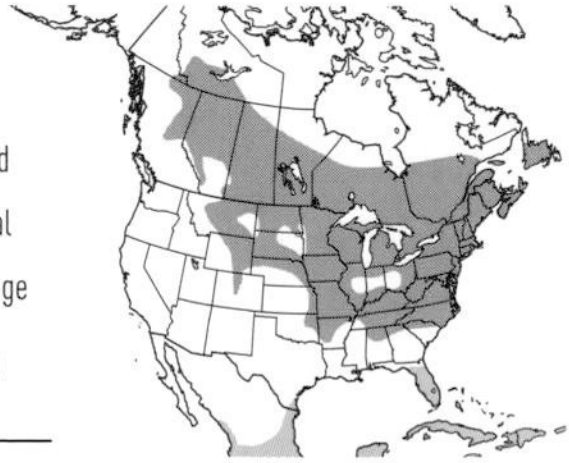

Location ______________________

Seiurus motacilla, Seiurus noveboracensis Length: 6"

WOW!
Lots of experienced birders struggle to tell the waterthrushes apart. Fortunately, the two species have very different songs.

LOOK FOR: Our two waterthrushes are confusingly similar, and both look more like tiny thrushes than warblers. For both, the constant tail bobbing as they walk is an excellent field mark. Males and females are similar, and they do not change appearance seasonally.

LISTEN FOR: The distinctive song of the Louisiana Waterthrush starts out with two or three down-slurred notes and ends in a sputtery jumble: *tee-yew, tee-yew, tee-yew, chicky-chick-a-chur-wow-chik!* Also gives a loud and sharp call note—*chink!*

REMEMBER: To determine *which* waterthrush you may be watching, examine the head. The Louisiana has a bold white eye line and plain unstreaked throat. The Northern has a finely streaked throat and narrow, often buffy (not white) eye line.

▶ *The Louisiana Waterthrush fearlessly wades in swift-running streams, tossing leaves and rocks to find insects and crustaceans.*

FIND IT: The Louisiana Waterthrush spends spring and summer along small flowing streams and ponds throughout the eastern U.S. Breeding territories are formed in a narrow corridor along streams.

LOUISIANA WATERTHRUSHES

☐ Date seen ____________

Location ____________

NORTHERN WATERTHRUSHES

☐ Date seen ____________

____

PROTHONOTARY WARBLER

Protonotaria citrea Length: 5½"

LOOK FOR: The Prothonotary Warbler (*pro-THON-oh-tary*) is a bird of wooded swamps, where its loud song rings out. The male's bright golden yellow head and breast contrast sharply with the large black bill and eyes, blue-gray wings, and white undertail. Females are duller overall.

LISTEN FOR: A clear, loud song on a single tone: *sweet-sweet-sweet-sweet-sweet!* Call note is a metallic *tchit!*

REMEMBER: The plain yellow face and head and large bill are good field marks for identifying the Prothonotary Warbler. Fall and winter birds may show a paler (not black) bill.

WOW!

The Prothonotary Warbler is named for the golden hood traditionally worn by the notary officer of the Roman Catholic Church.

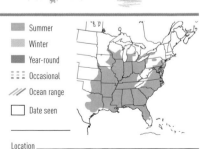

▶ *A male Prothonotary Warbler wintering in a mangrove swamp.*

FIND IT: In its swampy woodland habitat, the Prothonotary Warbler is often heard before it is seen. Males will sing from the treetops, but both sexes will forage quite low. Nests in tree cavities or nest boxes in trees standing in water.

Summer
Winter
Year-round
Occasional
Ocean range
Date seen ________

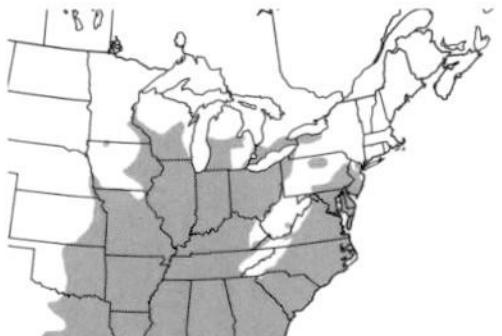

Location ________

Geothlypis trichas Length: 5"

LOOK FOR: With their black masks and the bright yellow throats for which they are named, adult males are easy to identify. Adult females and young birds are less clearly marked, but still offer clues in the obvious yellow throat patch, darkish face, and plain tan (not yellow) belly.

LISTEN FOR: Yellowthroats are very vocal birds, singing a clear, ringing *witchety, witchety, witchety.* Also utters several scold notes and rattles. Most common call note is buzzy and nasal: *cherk!*

REMEMBER: Several other warbler species might be confused with the Common Yellowthroat, but no other warbler shows this exact pattern.

WOW!

One of the Common Yellowthroat's folk names, Black-masked Ground Warbler, reflects the yellowthroat's tendency to stay low to the ground nearly all the time.

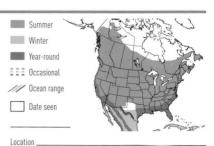

◀ *Unlike most warblers, Common Yellowthroats prefer brushy habitat to deep woods.*

FIND IT: Our only warbler that prefers to nest in marshy habitat, the Common Yellowthroat can be hard to see because it prefers deep, brambly cover. Pish a few times, and it may get curious enough to pop into the open for a look at you.

Summer
Winter
Year-round
Occasional
Ocean range
Date seen ______

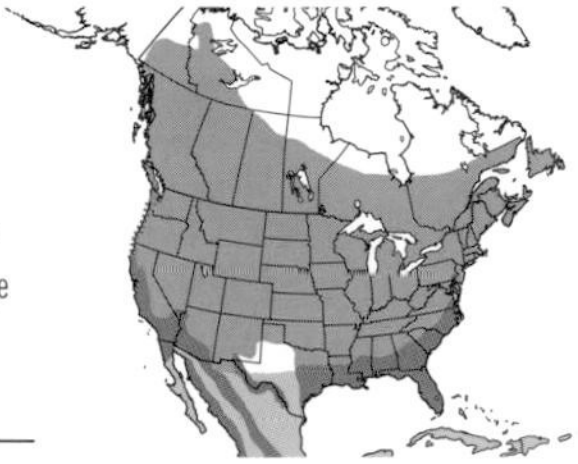

Location ______

Wilsonia pusilla Length: 4¾"

LOOK FOR: Adult male's round black cap, plain yellow face, and unmarked yellow underparts are diagnostic field marks. Females have yellow-olive cap, yellow eyebrow, and plain yellow face. Back is olive-yellow in both sexes and tail is dark and unmarked.

LISTEN FOR: Song is a loud, ringing series of harsh chips dropping in tone and speed toward the end: *chi-chi-chi-chi-CHETCHETCHET.* Call note is a loud *CHET!*

REMEMBER: Wilson's Warblers have a "beady-eyed" look that few other warblers show. Yellow Warbler is similar but bright yellow overall, including the back.

WOW!

The Wilson's Warbler is named for Alexander Wilson (1766–1813), one of North America's first ornithologists.

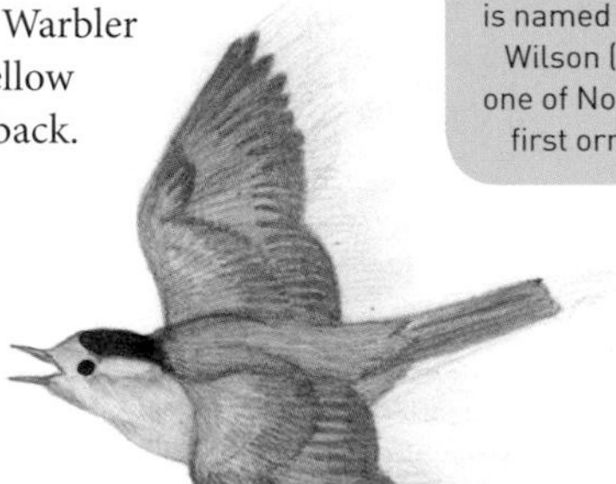

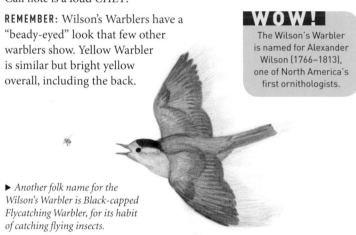

▸ *Another folk name for the Wilson's Warbler is Black-capped Flycatching Warbler, for its habit of catching flying insects.*

FIND IT: More common in the West than East. Common low to ground in thick, brushy woods and alder and willow thickets, especially along streams and near other water. Abundant spring migrant in the West.

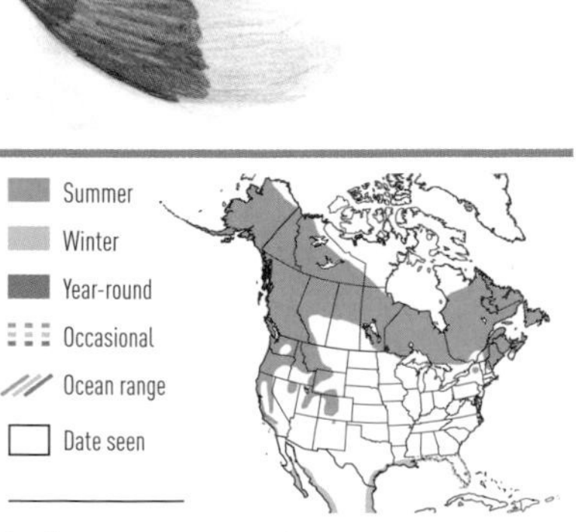

Date seen ______________

Location ______________________________

Icteria virens Length: 7"

WOW!

In his flight display, a male Yellow-breasted Chat flies across an open space, calling loudly and flapping his wings awkwardly like a wind-up toy that is freaking out.

LOOK FOR: The Yellow-breasted Chat is our largest warbler, though it does not really look, act, or sound like a warbler. It has a large head, a stout black bill, and a long tail. Sexes are similar, and plumage does not vary by season.

LISTEN FOR: Chats make a huge variety of weird noises, including whistles, clucks, chucks, and harsh scolding notes delivered in a series, each call separate and distinct. It sounds like this: *cherrk eeeep! woo-woo-woo! chek! wok! ank-ank-ank!*

REMEMBER: The chat's size and bright yellow throat should be enough to identify this species. No other songbird has the clear yellow throat and breast and white spectacles (connected eye-rings).

► *A Yellow-breasted Chat launches into a flight song by the light of a full moon. Chats are insomniacs in spring.*

FIND IT: Yellow-breasted Chats often sing from the deep cover of brushy tangles along woodland edges, old overgrown fields, and streamside thickets, but they can be coaxed out by pishing or by imitating their odd sounds.

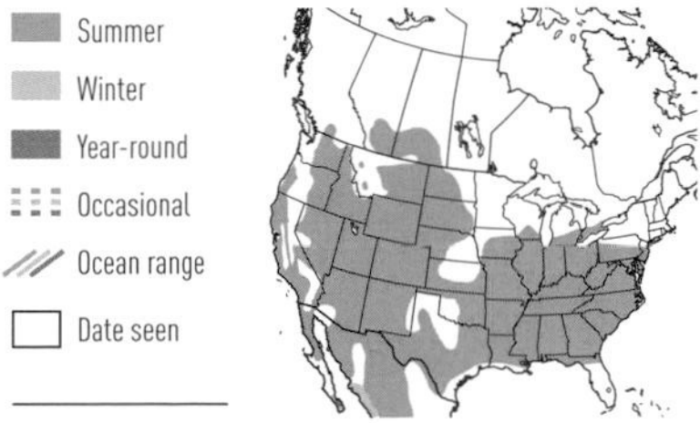

Location ____________

Piranga rubra Length: 7¾"

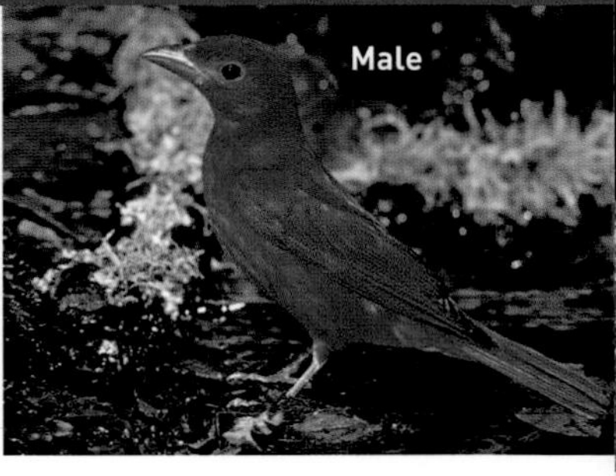

LOOK FOR: The male Summer Tanager is a bright rosy red all over and can at times appear to have a slight crest on its head. The other most noticeable feature of this species is its large dark bill, much larger than the bill of the Scarlet Tanager. The yellow-green females may show a hint of the male's reddish coloration.

WOW!
One of the folk names for this species is Beebird, for its habit of eating bees and wasps. Its long stout bill is an ideal tool for capturing and subduing these stinging insects.

LISTEN FOR: Song is similar to the Scarlet Tanager's, but richer—less harsh and more liquid. More distinctive is the Summer Tanager's explosive call: *perky-tuck!* or *perky-tucky-tuck!*

REMEMBER: Female Summer Tanagers are more uniformly yellow than female Scarlet Tanagers. If you're confused, look at the bill. If it's long and thick at the base, you've got a Summer Tanager.

◀ *The Summer Tanager's strong, toothed bill is more than a match for the sting of a yellow jacket.*

FIND IT: In summer, look for Summer Tanagers in open oak or pine woods in the Southeast and in streamside cottonwoods in the Southwest. Finding one in the treetops can take patience, as they are not super active when foraging.

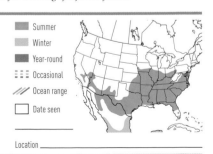

Summer
Winter
Year-round
Occasional
Ocean range
Date seen ______

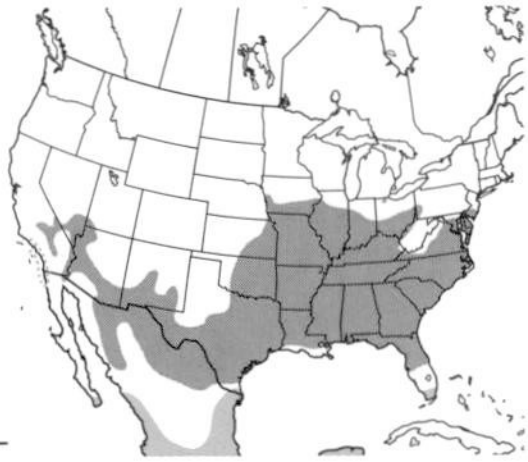

Location ______

LOOK FOR: The spring and summer plumage of the male Scarlet Tanager is stunning and unmistakable. In fall, males lose all their red color and turn a dull yellow-olive (though they retain the black wings and tails). Females are olive, yellow, and gray overall in all seasons.

LISTEN FOR: A Scarlet Tanager sounds like an American Robin singing with a sore throat: *cheer-ree, chee-rear, cheer-ree, cheer-wow!* Another commonly given call is *chip-burr!*

REMEMBER: The Western Tanager and the female Summer Tanager look similar to the female Scarlet Tanager, but their backs are darker in the center.

WOW!

No other bird in North America has the male Scarlet Tanager's combination of scarlet red body and black wings. It's too bad he has to molt into his duller nonbreeding plumage each fall.

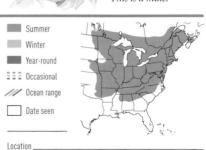

◀ *Both adult Scarlet Tanagers feed nestlings. This is a male.*

FIND IT: Common in leafy eastern woodlands, Scarlet Tanagers seem to prefer oak-dominated forests with large trees. Despite the male's bright colors, Scarlet Tanagers can be difficult to spot in thick foliage high in the treetops.

Summer
Winter
Year-round
Occasional
Ocean range
Date seen ______

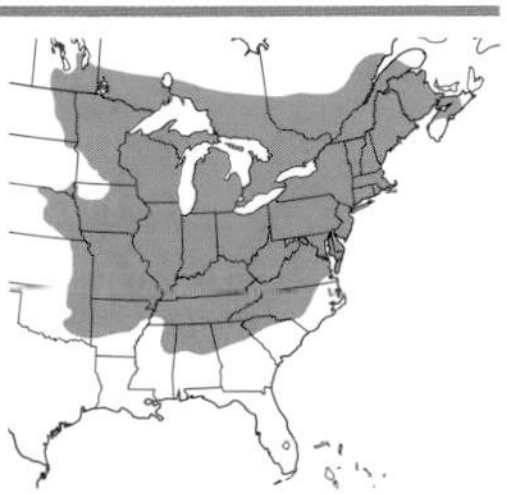

Location ______

WESTERN TANAGER

Piranga ludoviciana Length: 7¼"

LOOK FOR: Male in breeding plumage is unmistakable: red head, lemon yellow body, black back and wings, with both yellow and white wing bars and yellow rump. Female is similar to other tanagers but shows a gray back and obvious wing bars.

LISTEN FOR: Short, hoarse, whistled phrases that are very robinlike in quality: *brr-eet, burry, burry, brr-eet!* Calls include a soft, rising *wheee?* and a percussive *pretty-tick!*

REMEMBER: This is our only common tanager with bold wing bars. The stout bill helps separate the Western Tanager from similar oriole species.

◀ *Western Tanagers spend a lot of time in the tops of trees, especially territorial adult males, singing to claim their territories.*

WOW!

The Western Tanager nests farther north than any of our other tanagers.

FIND IT: A treetop specialist that prefers open pine or mixed forest during the breeding season, often at higher elevations. Visits many other habitats in migration. Less active than some songbird species, so familiarity with the song is useful in finding the Western Tanager.

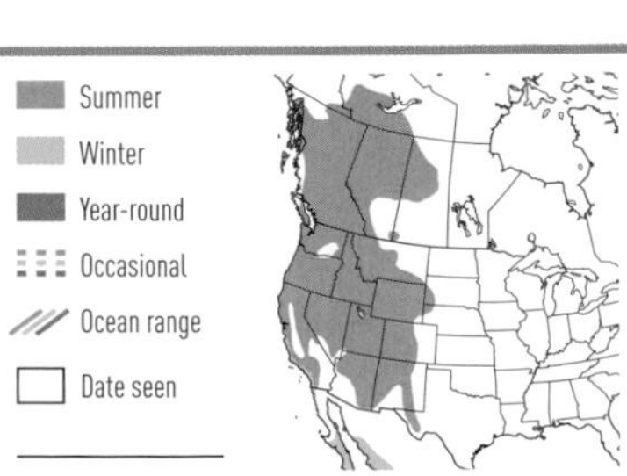

____

Location ______________

GREEN-TAILED TOWHEE

Pipilo chlorurus Length: 7¼"

LOOK FOR: Named for its olive green upperparts and tail, the Green-tailed Towhee's most noticeable features are its rusty cap and white throat, contrasting with the gray head and breast. This is our smallest towhee.

LISTEN FOR: Song is loud and ringing: *chert, wee-wee, CHEEE churr.* Song often compared to song of Fox Sparrow. Call is cat-like: *myeeew.*

WOW!
Singing male Green-tailed Towhees may incorporate the songs of other nearby species into their own songs. Copycats!

REMEMBER: This ground-loving bird is curious and can be lured into view by spishing. But when alarmed it may run away rather than fly.

▼ *If you don't hear the Green-tailed Towhee's call or song first, you might hear the noise created as a towhee scratches through leaf litter while foraging.*

FIND IT: Common in a variety of dry, brushy habitats in the western mountains, including chaparral, sagebrush, riparian woods, and thickets. Often heard before it is seen.

Summer
Winter
Year-round
Occasional
Ocean range
☐ Date seen

____

Location ______________

EASTERN TOWHEE and SPOTTED TOWHEE

Pipilo erythrophthalmus, Pipilo maculatus Length: 8"

LOOK FOR: The former name of this species described the bird better: Rufous-sided Towhee, named for the rusty rufous sides. Spotted: wings and back of both male and female are spotted with white. Longer tailed and larger than our other sparrow species.

LISTEN FOR: One of the easiest bird songs to learn and remember, the Eastern Towee sings *drink your teeeaa!* Or simply *drink teeeaa!* Call is a similar-sounding *chew-ink!* They also use a buzzy, rising *tzeeeee!* as a flight call and alarm call. Spotted: song is *sweet-sweet teeeeaaa!* But it's less musical than Eastern's song.

REMEMBER: Both Eastern and Spotted Towhees are larger and more colorful than the Dark-eyed Junco, which also has a black hood and back.

▼ *The Eastern Towhee leaps forward, kicking leaf litter back. In this way, it uncovers hidden seeds and insects.*

WOW!

These two towhee species were once considered a single species: Rufous-sided Towhee.

FIND IT: Eastern: common in thick undergrowth and along brambly woodland edges and old-fields. Spotted: in willows, sagebrush, chaparral, and brushy woods. Both will visit backyard feeders near appropriate habitat for seed bits scattered on the ground.

EASTERN TOWHEE

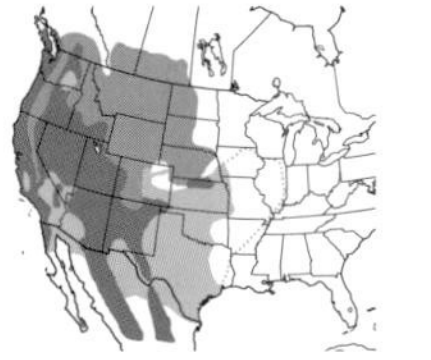

WESTERN TOWHEE

☐ Date seen ____________

☐ Date seen ____________

Location ____________ ____________

CANYON TOWHEE and CALIFORNIA TOWHEE

Pipilo fuscus, Pipilo crissalis Length: 8¾", 9"

LOOK FOR: Both of these plain brownish towhees look like what they are: giant sparrows. Canyon Towhee has a faint rusty cap and is slightly paler overall than California Towhee, which shows rust on the throat. California is plain breasted and dark bellied. Canyon usually shows a spotty necklace, central breast spot, and whitish belly.

WOW!

These two species were once lumped together as the Brown Towhee, which is a descriptive name but kind of boring compared to Canyon and California.

LISTEN FOR: Canyon: song is a loud and ringing series of notes: *chee-chee-chee chee-chee-chee-chee*. Call is a woodpecker-like, nasal *kyerr.* California: song starts with metallic chips, speeds up, then drops in tone: *chip-chip-chichichichi-drrdrrdrr.* Call: metallic *chip!*

REMEMBER: These species can be told apart by range and by song. Canyon can be distinguished from similar Abert's Towhee (not shown) by Abert's partially black face.

◀ *Being ground-loving birds, Canyon (shown) and California Towhees run away from danger rather than fly.*

FIND IT: Both species prefer open arid scrub with scattered brush. Canyon Towhee inhabits the desert Southwest east of California. California Towhee is found in California and southwestern Oregon.

Summer
Winter
Year-round
Occasional
Ocean range

☐ Date seen ______________ ☐ Date seen ______________

Location ______________ ______________

LARK SPARROW

Chondestes grammacus Length: 6½"

LOOK FOR: This large sparrow's bold facial markings of rufous, black, and white are diagnostic. Also note the dark central breast spot on white underparts. White sides and tip of dark tail are especially obvious in flight.

LISTEN FOR: Song is a rich, musical blend of whistles, trills, and buzzes and is fairly long in duration. Call is *tsink* and is often given in flight.

REMEMBER: No other North American sparrow has the Lark Sparrow's combination of a bold, colorful facial pattern and central breast spot.

WOW!

Male Lark Sparrows are known to sing at night.

▶ *Lark Sparrows love to hang out with their friends in small flocks while foraging.*

FIND IT: A sparrow of open country with scattered bushes, trees, and fencerows; plus farmyards, pastures, and grassland with bare ground. Often forages on the ground in the open. Nearly always in small flocks, especially in migration and winter.

Summer

Winter

Year-round

Occasional

Ocean range

Date seen ____________

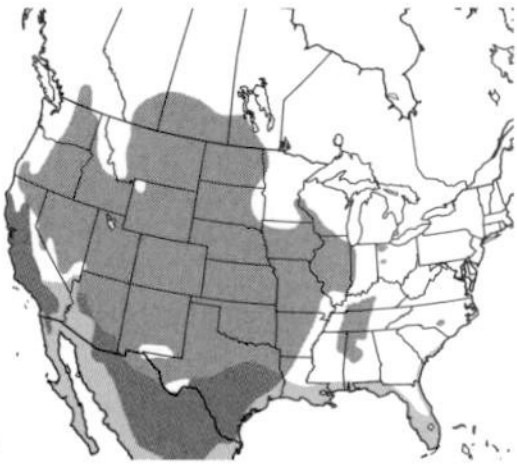

Location ____________

CHIPPING SPARROW

Spizella passerina Length: 5½"

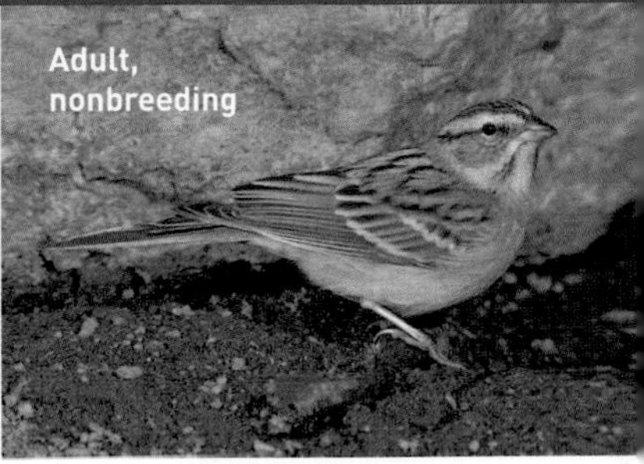

LOOK FOR: In breeding plumage the adult has a rusty cap over a white eyebrow and a black eye line. Adults in nonbreeding plumage look like faded versions of their summer selves. First-year birds can be confusing, with finely streaked heads and pale faces. All Chippies look flat headed and have a grayish rump.

WOW!

Chipping Sparrows love to line their nests with hair. Place some hair from your next haircut on your lawn in spring, and in the fall you may find a nest with familiar-looking hair in it.

LISTEN FOR: Chipping Sparrows sing a long dry trill on a single note that sounds like a cross between an insect and a sewing machine. They also utter single *chips*, for which they are named.

REMEMBER: Though the Clay-colored Sparrow looks similar, it prefers brushy, treeless areas. Clay-coloreds have brownish tan rumps instead of gray.

▶ *Chipping Sparrows often nest low in landscape shrubbery near buildings.*

FIND IT: Chipping Sparrows prefer open woodlands, wooded parks, and even suburban neighborhoods, where they will nest in the most landscaped of habitat. They come to bird feeders for mixed seed and cracked corn.

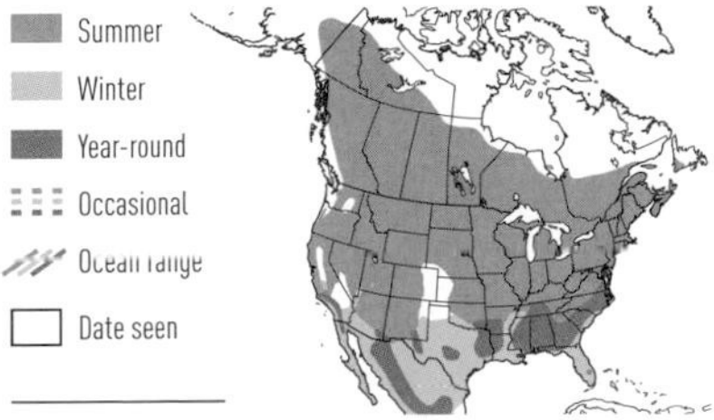

Date seen ______

Location ______

CLAY-COLORED SPARROW

Spizella pallida Length: 5½"

LOOK FOR: Breeding adult has a dark crown, white eye stripe, and a subtle dark mustache. Very similar to a nonbreeding-plumaged Chipping Sparrow. The clear gray nape and brownish rump on the Clay-colored help separate the two species.

LISTEN FOR: One of the least musical of all sparrow songs: a series of nasal buzzes: *tzeee-tzeee-tzeee.* Call is a soft *tsip.*

REMEMBER: Clay-colored Sparrows have a slate gray nape and a brownish rump. Nonbreeding Chipping Sparrows have a grayish nape and a grayish rump.

WOW!

Male Clay-colored Sparrows defend their breeding turf vigorously, excluding both Song and Chipping Sparrows where their ranges overlap.

◀ *The Clay-colored Sparrow is a frequent victim of nest parasitism by the Brown-headed Cowbird.*

FIND IT: Common in summer grassland and prairie habitat with scattered shrubs or bushes. Males perch on top of shrubs and sing their insectlike song. Often heard before they are seen, but easy to find thereafter.

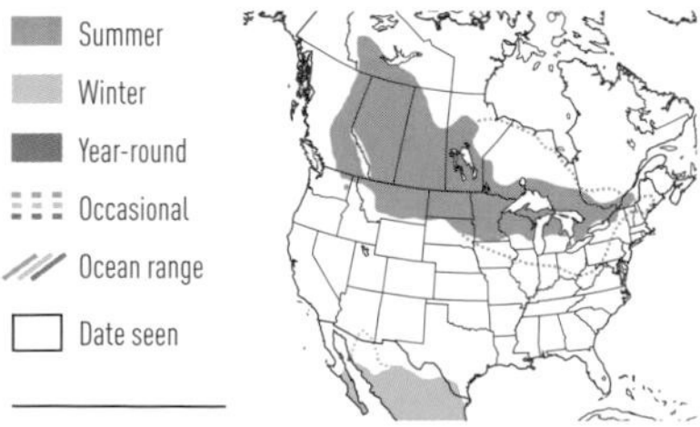

Location ______________________

LOOK FOR: All the Field Sparrow's distinctive field marks are on the head and breast. The plain-faced look is emphasized by a white ring around the dark eyes. Adults are similar and do not change plumage seasonally.

LISTEN FOR: A series of sweet, whistled notes that speeds up into a trill and drops in tone toward the end: *too-too-too-too-tootootoottititititititi.* The pattern is similar to the rhythm of a dropped Ping-Pong ball, which bounces faster until it stops.

REMEMBER: The Field Sparrow's pink bill, plain face (with white eye-ring), and unmarked breast help to separate this species from the similar but less common American Tree Sparrow.

▼ *It's safer for Field Sparrow chicks to hide separately in deep cover than to huddle together in the nest, where their scent could attract a predator.*

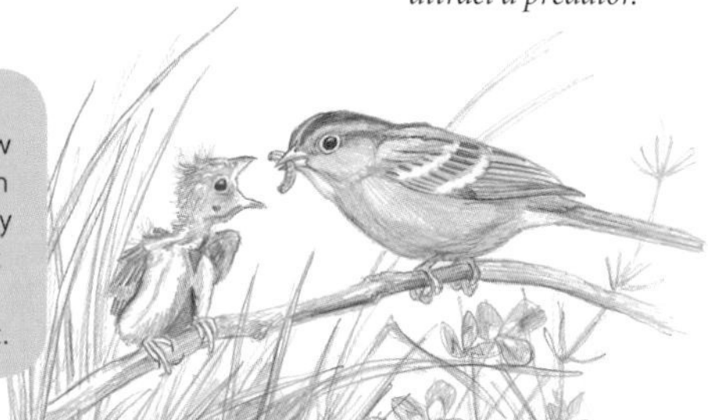

WOW!

The young Field Sparrow looks very different from its parents. It can be very confusing to identify, at least until one of the parents comes to feed it.

FIND IT: In spring and summer, males will sing while perched on a weed stem or sapling above the grass of old-fields (scrubby overgrown meadows). Field Sparrows will visit bird feeders in rural settings for mixed seed and cracked corn.

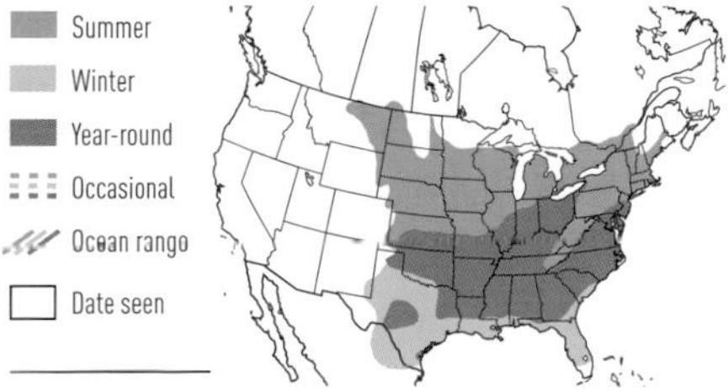

Date seen ____________

Location ____________

LOOK FOR: This large, boldly marked bird is named for its fox red coloration, but not all Fox Sparrows are reddish—some western birds are dark brown or even gray. Sexes are similar, and coloration does not vary seasonally.

LISTEN FOR: Fox Sparrows have a beautiful song that starts with several sweet whistled notes, then a trill, and ends with a short warble: *sweet-sweet, chee-chee-chee-tititititi-chew-wee!* Call note is a loud *chip!* or *smak!*

REMEMBER: Though similar in size, Fox Sparrows appear stockier than Song Sparrows. The fox red coloration is brighter than the Song Sparrow's earth tones, and the breast of the Fox Sparrow is spotted with rusty brown (the Song Sparrow's breast is streaked with brown).

WOW!

The various forms of the Fox Sparrow (red, gray, sooty, and large-billed) appear very different and may one day be split into separate species.

▶ *A Fox Sparrow uncovers seeds and insects by grabbing leaf litter with its toes in a two-footed shuffle and kicking it back behind.*

FIND IT: Fox Sparrows prefer to forage on the ground, often in dense, brushy cover, scratching for food like a towhee. They do not normally winter in flocks, but they will join other birds in a thicket or feeding below a bird feeder.

Date seen ____________

Location ____________

SONG SPARROW

Melospiza melodia Length: 6¼"

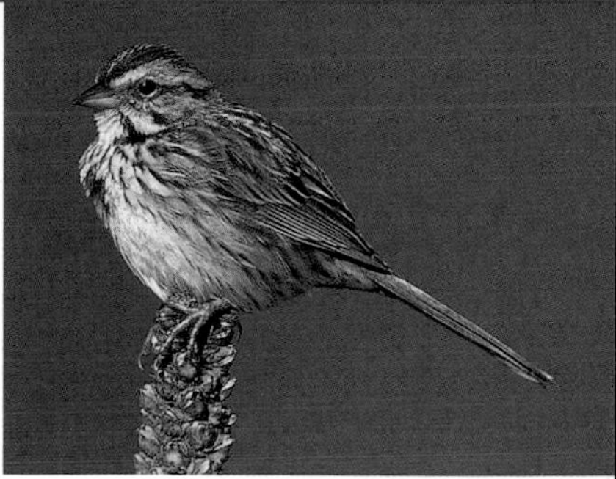

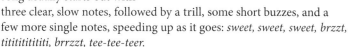

LOOK FOR: The chunky, streaky Song Sparrow's best-known field mark is the central breast spot, where the breast streaks come together to form a noticeable splotch. Its heavy streaking overall gives it a dark, dusky look.

LISTEN FOR: The highly variable song usually starts out with three clear, slow notes, followed by a trill, some short buzzes, and a few more single notes, speeding up as it goes: *sweet, sweet, sweet, brzzt, tititititititi, brrzzt, tee-tee-teer.*

REMEMBER: Learn the Song Sparrow's field marks well. Then, when you encounter an unfamiliar sparrow, ask yourself, "What makes this bird different from a Song Sparrow?" The answers will be the new bird's important field marks.

WOW!

Not all Song Sparrows look alike. There may be as many as 30 different forms of this species across the continent. Fortunately, all these Song Sparrows sing the same beautiful song.

► *Song Sparrows are very sneaky and secretive near their nests.*

FIND IT: Our most widespread sparrow in North America, the Song Sparrow prefers dense cover such as brushy field edges, hedgerows, and brambles, but it's also common in backyards, parks, and cemeteries.

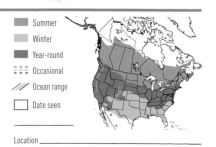

Summer
Winter
Year-round
Occasional
Ocean range
Date seen

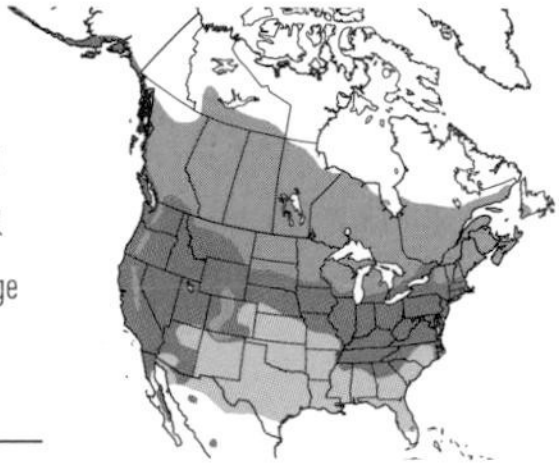

____

Location __________

LINCOLN'S SPARROW

Melospiza lincolnii Length: 5¾"

LOOK FOR: A warm, colorful sparrow similar to Song Sparrow, but smaller and more finely marked. Note fine streaks on breast, gray eyebrow, buffy mustache, and buffy eye-ring.

LISTEN FOR: Song is a bright, clear jumble of phrases, rising in volume, pitch, and intensity in the middle: *burrburr ZEEEEEEEE chrup.* Sometimes compared to Purple Finch and House Wren. Calls include *chup* and *zeeet.*

REMEMBER: The Lincoln's Sparrow looks like a Song Sparrow that got a makeover.

WOW!

This species is not named for Abe Lincoln; it's named for Thomas Lincoln, who accompanied John James Audubon on a bird-finding trip in 1833.

◄ *Where the two species' ranges overlap, the Song Sparrow outcompetes the Lincoln's Sparrow.*

FIND IT: Nests in thickets of alder and willow and in brushy habitat near water. In winter found in brushy woodland-edge habitat, usually near water. The Lincoln's Sparrow is a shy bird and often goes undetected.

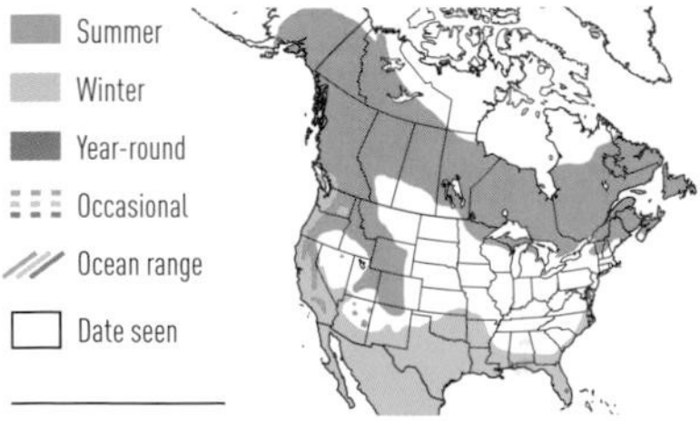

Location ______

WHITE-THROATED SPARROW

Zonotrichia albicollis Length: 6¾"

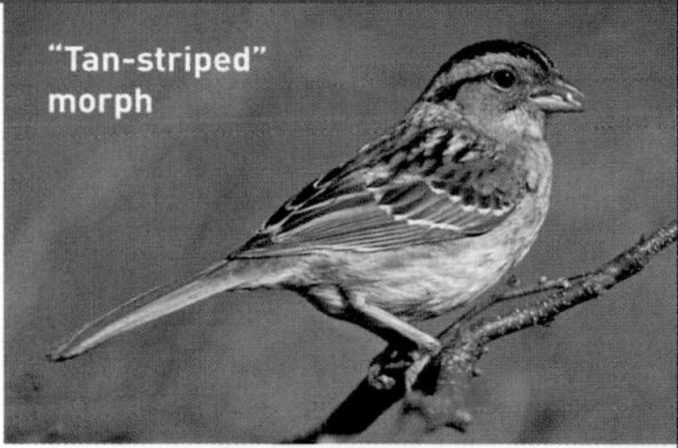

LOOK FOR: One of several species of crowned sparrows in North America, the White-throated Sparrow comes in two color varieties, or morphs: birds with white-striped crowns and birds with tan-striped crowns. Sexes are similar, and coloration does not change seasonally.

LISTEN FOR: *Old sam peabody, peabody, peabody* is the sweetly whistled, almost sad-sounding song. Two call notes are *tseeet!* and a loud, metallic *chink!*

REMEMBER: The White-throated Sparrow's white throat sets it apart from the similar but larger White-crowned Sparrow, which has a boldly striped head but a plain gray throat.

◀ *White-throated Sparrows will sing at almost any time of year, even in winter if it's sunny.*

WOW!

It was once thought that the tan-striped morph of this species was the juvenal plumage of the white-striped adults. Studies have shown that white-striped adults usually mate with tan-striped birds.

FIND IT: The White-throated Sparrow prefers scrubby undergrowth along woodland edges and weedy fields but visits bird feeders for cracked corn, mixed seed, and sunflower seed bits.

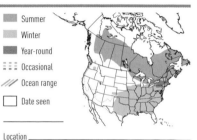

Summer

Winter

Year-round

Occasional

Ocean range

☐ Date seen

____

Location ________________________

WHITE-CROWNED SPARROW

Zonotrichia leucophrys Length: 7"

WOW!

In a 1962 experiment, several hundred White-crowned Sparrows were trapped in California and released in Maryland. One year later, eight of them had found their way back.

LOOK FOR: The White-crowned Sparrow's plain gray throat and unmarked breast help set it apart from other sparrows. Sexes are similar, and color of adults does not change seasonally. First-winter birds have brown-and-gray-striped heads until the following spring.

LISTEN FOR: White-crowned Sparrows start their song with one or two clear notes followed by a series of burry, buzzy warbles that rise up the scale: *you-can-be-so-cheez-ee!* Call note is a bright *seep!*

REMEMBER: Your first impression of the White-crowned Sparrow may be of a large gray sparrow with an erect posture. One look at the bold head stripes and clear gray throat, and you'll know it can only be a White-crowned Sparrow. Don't forget about the young birds' tan-striped heads!

► *An adult White-crowned Sparrow forages for seeds on a stalk of lamb's quarters.*

FIND IT: In the East the White-crowned Sparrow is most often seen in winter and during migration. Preferred winter habitat is in scrubby hedgerows along fields, woodland edges, and thickets, where it can be found in small flocks.

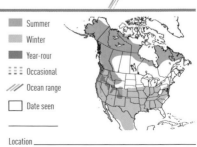

Summer
Winter
Year-rour
Occasional
Ocean range
☐ Date seen

____

Location ______________

GOLDEN-CROWNED SPARROW

Zonotrichia atricapilla Length: 7¼"

LOOK FOR: Adult Golden-crowned Sparrow has a golden crown stripe surrounded by black. Slightly larger than, but very similar to, the White-crowned Sparrow. Immature birds may have only a hint of the yellow crown and lack the black head stripes.

WOW!

In a mixed winter flock of sparrows in the West, the Golden-crowns are more likely than other species to stick to the thicker underbrush.

LISTEN FOR: Song is a sad-sounding whistle: *oh dear meeee!* Or *oh lonely meee!* Call is a clear, soft *chew!*

REMEMBER: Subadult Golden-crowned Sparrows can be identified by their plain gray face and grayish bill. Otherwise they are very similar to young White-crowned Sparrows.

◄ *The male Golden-crowned Sparrow is known to bring food to his mate while she is incubating eggs.*

FIND IT: Nests in stunted boreal woods and tundra scrub as far north as the tree line. Winters in dense thickets and underbrush, chaparral, parks, and gardens as far south as the Mexico border. Often found in mixed flocks with White-crowned Sparrows. Will visit feeding stations for seed scattered on the ground.

Summer

Winter

Year-round

Occasional

Ocean range

Date seen ______

Location ______

HARRIS'S SPARROW

Zonotrichia querula Length: 7½"

LOOK FOR: Harris's Sparrow is our largest sparrow, and its combination of a black face and pink bill (in adult birds) makes it relatively easy to identify. First-winter birds have a pale brown head and a white throat.

LISTEN FOR: If you want to hear a Harris's Sparrow sing, you'll have to head north to its breeding habitat (see below). Song is two or three clear whistles followed by two more on a lower tone: *see-see-see* [pause] *soo-sooo.* Call is *check.*

REMEMBER: The Harris's Sparrow's large size is a great clue to its identity, especially when compared with less boldly marked young birds.

▼ *Feeder watchers in the Midwest feel extra lucky when a Harris's Sparrow visits their feeders in the winter.*

WOW!

The Harris's Sparrow was named by John James Audubon for Edward Harris, his friend and travel companion in the mid-1840s.

FIND IT: Uncommon in winter in brushy areas, hedgerows, and open woodland. Summers in far North in boreal forest and tundra scrub. May mix with White-crowned Sparrows in winter feeding flocks. Sometimes visits feeders in the Midwest.

Summer
Winter
Year-round
Occasional
Ocean range
Date seen ________

Location ________

DARK-EYED JUNCO

Junco hyemalis Length: 6¼"

LOOK FOR: The Dark-eyed Junco is a well-known bird even among nonbirders. Several different forms exist throughout the West, and all were once considered separate species. In flight, all Dark-eyed Juncos show obvious white outer tail feathers.

WOW!

The folk name of Snowbird is for their winter-weather coloration (gray skies above, snow on the ground) and because they appear at feeders at the first snowfall.

LISTEN FOR: Juncos are members of the sparrow family, and they sing a long ringing trill—*tiitiitiitiitiitii*—that sounds similar to a Chipping Sparrow's song but is slower and more musical. They also give a number of chip notes (*tick!*) and short buzzes in all seasons.

REMEMBER: East of the Great Plains, the slate-colored Dark-eyed Junco is most common. In the West, five other forms of the species are present.

▶ *Flocking Dark-eyed Juncos bicker and show aggressive behavior often, their angry twitters sounding like skates on ice.*

FIND IT: Flocks of juncos forage along woodland edges and hedgerows, hopping over the ground in search of seeds, often chipping at each other. At bird feeders they eat cracked corn and mixed seed.

Summer

Winter

Year-round

Occasional

Ocean range

Date seen

Location

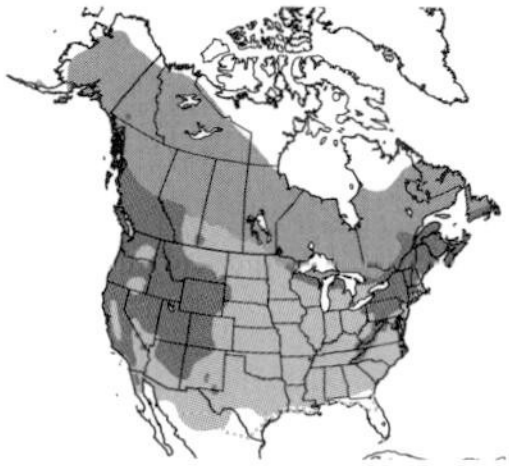

LAPLAND LONGSPUR

Calcarius lapponicus Length: 6"

Winter

Breeding male

LOOK FOR: Few of us get to see the male Lapland Longspur in his full breeding glory, with his black face outlined in white and the chestnut nape and collar. Winter male retains a partial collar of rust, a tan cheek outlined in dark, and heavy dark streaks on the sides. All plumages show narrow white sides to the tail in flight.

LISTEN FOR: Song in display flight is a rich, throaty warble, similar in tone to Bobolink. More commonly heard are calls: a whistled *tyew* and a dry, buzzy note.

REMEMBER: Though it nests the farthest north, in winter this is our most common longspur species and the easiest to add to your life list.

▼ *Lapland Longspurs are often found in mixed winter flocks with Horned Larks and Snow Buntings.*

WOW!

Lapland Longspurs returning to the Arctic tundra to breed retain extra body fat to help them survive late-spring cold and snow.

FIND IT: An abundant breeding bird on the tundra and wet meadows of the far North. In winter, found in open grassland, plowed or planted fields, prairies, and shorelines. Large winter flocks may number in the thousands.

Summer
Winter
Year-round
Occasional
Ocean range
Date seen

Location ____________

SNOW BUNTING

Plectrophenax nivalis Length: 6¾"

LOOK FOR: Most of us don't get to see the male Snow Bunting in its snowiest attire—its pure white-and-black breeding plumage worn when the birds are on the Arctic tundra. In winter, when they are present across much of northern North America, both males and females are rusty and white with black wingtips.

WOW!

Snow Buntings can handle extreme cold—as low as 58 degrees below zero. They may burrow into the snow to shelter from the freezing wind and stay warm.

LISTEN FOR: The Snow Bunting's calls are a whistled descending *cheew!* and a harsh, nasal *brzzzt!* given in winter flocks. The song is a musical warble full of *tyeew* notes.

REMEMBER: Snow Buntings like to mix with larks and longspurs. When a mixed flock takes flight, the birds flashing white wings with black tips are Snow Buntings.

▶ *Snow Buntings flock in dunes and on flats along the coasts in winter, their white wings twinkling when they take flight.*

FIND IT: Summers are spent in the far North on the tundra. In winter, flocks move south, where they prefer open areas such as beaches and farm fields (especially ones with freshly spread manure).

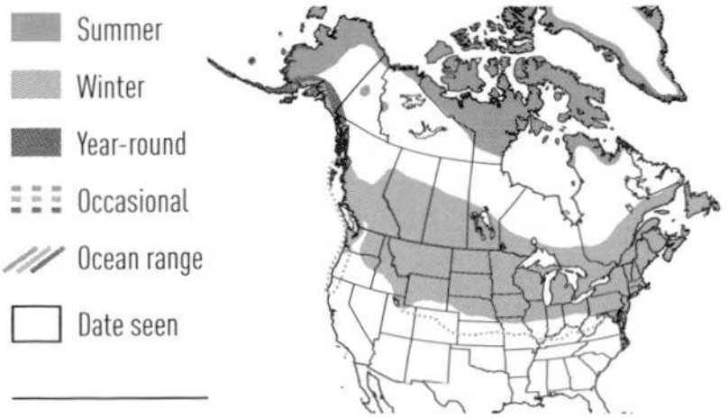

Location ____________________

ROSE-BREASTED GROSBEAK

Pheucticus ludovicianus Length: 8"

LOOK FOR: The Rose-breasted Grosbeak is a big-headed, large-billed bird (its name means "large bill") of the treetops. Females and young males are streaky. The fall and winter adult male wears a faded, splotchy version of the breeding plumage.

WOW!

In spring migration, Rose-breasted Grosbeaks often show up at bird feeders where sunflower seed is offered, sometimes prompting a "Wow! What's that bird?"

LISTEN FOR: An American Robin that's had singing lessons describes the rich, musical warble: *chewee-churweo, chewee-turleo, chewee-churweo.* Call note is a sharp squeak: *eek!*

REMEMBER: Female Rose-breasted Grosbeak and female Purple Finches look similar. The grosbeak is much larger overall and larger headed, with a broad white eye line, a finely streaked breast, and a pale, pinkish bill. The finch has a messy eye line, broadly streaked breast, and a gray bill.

▶ *A male Rose-breasted Grosbeak in flight is a pinwheel of black, white, rose, and carmine.*

FIND IT: Prefers young, open deciduous woods during spring and summer. Often forages in thick foliage near the treetops and can be best located by song or call note. In migration, it can appear almost anywhere.

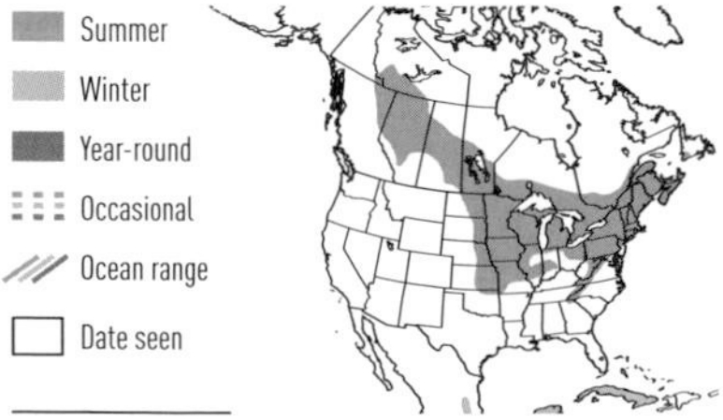

Location ______

BLACK-HEADED GROSBEAK

Pheucticus melanocephalus Length: 8¼"

LOOK FOR: Breeding-plumaged adult male has a black head and orange collar, underparts, and rump. Black wings are boldly marked with white. Black back shows vertical orange stripes. Adult female is butterscotch colored below with finely streaked sides. Face is boldly patterned with dark and light stripes. Stout bill is dark above, light gray below.

WOW!

The Black-headed Grosbeak is one of the few bird species that will eat monarch butterflies, which are toxic to most birds.

LISTEN FOR: Song is a robinlike series of warbled and whistled phrases that rise and fall in tone. *Up here! Way up here! I'm singing this beautiful song for you!* Call is a sharp, spitting *tick!*

REMEMBER: Where ranges overlap, the similar-looking female Rose-breasted Grosbeak can be identified by her heavily streaked breast. Female Black-headeds have a warmer-colored breast unstreaked in the center.

◀ *Male Black-headed Grosbeaks perform impressive song flights above perched females during courtship.*

FIND IT: Spends summers in deciduous woods, especially oaks and mixed oak-pine forests. Also in cottonwoods and willow groves along streams. Usually found in upper levels of trees. Winters in the tropics.

Summer
Winter
Year-round
Occasional
Ocean range
☐ Date seen

____

Location ______________

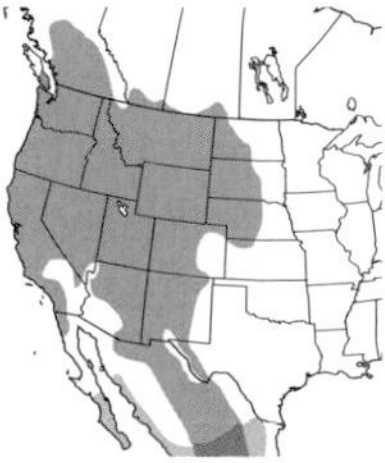

NORTHERN CARDINAL

Cardinalis cardinalis Length: 9"

WOW!

Oh no—a bald cardinal! When Northern Cardinals get infested with feather mites on the head, where they cannot preen, they can lose their head feathers. Once mite-free feathers grow in, the bird looks normal again.

LOOK FOR: The male Northern Cardinal is our only crested red bird, and it's hard to mistake it for another species. The male's black facemask and the massive seed-crushing reddish bill are easy to see. Females and juvenile birds are crested and reddish brown overall, but the female has a reddish bill and subtle black face while the young bird has a plain face and a black bill.

LISTEN FOR: Northern Cardinals have a loud, ringing song that can vary: *purty-purty-purty, what-cheer! what-cheer! tee-tee-tee-tee-tee!* Also gives a loud *pik!* call.

REMEMBER: Our red tanager species are similar in color but not in shape: they lack the cardinal's obvious crest.

► *A male Northern Cardinal sings and displays for a female, leaning from side to side and fluttering its wings.*

FIND IT: A southern bird that is expanding its range northward, the Northern Cardinal favors brushy habitat along woodland edges, open woods, parks, and backyards with dense cover, where it will visit bird feeders for sunflower seed.

Summer

Winter

Year-round

Occasional

Ocean range

Date seen

Location

PYRRHULOXIA

Cardinalis sinuatus Length: 8¾"

LOOK FOR: The "Desert Cardinal" looks like a Northern Cardinal that has gone gray. Male Pyrrhuloxia is light gray overall with a red face and a red central stripe from the chin to the lower belly. Long crest is tipped in red. Stubby, conical bill is pale yellow (cardinal bills are red). Female is gray-brown overall with a pale face and patches of red in wings, tail, and tip of slender crest.

WOW!

One of the other folk names for the Pyrrhuloxia is Parrot-billed Cardinal, for the bird's stubby and curved yellow bill.

LISTEN FOR: Song is very cardinal-like but higher pitched, sweeter, and more drawn out in tempo. Call is a metallic *chink*.

REMEMBER: It's pronounced *peer-uh-LOX-ee-uh*. Use this knowledge to impress your friends.

▶ *Pyrrhuloxias prefer thorny, brushy habitat—much drier habitat than that preferred by Northern Cardinals.*

FIND IT: A resident of brushy desert habitat: mesquite thickets, thorn scrub, streamside brush. Usually found in small flocks, which move to more open and wooded habitat in winter.

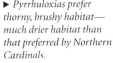

Summer
Winter
Year-round
Occasional
Ocean range
Date seen

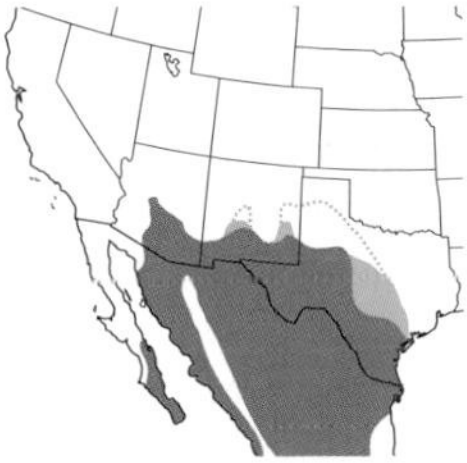

Location ______

BLUE GROSBEAK

Passerina caerulea Length: 6¾"

LOOK FOR: The Blue Grosbeak looks like a supersized Indigo Bunting with ketchup on its shoulders. Blue Grosbeaks regularly flick and spread their tails—a good field mark when light conditions obscure the birds' color (they can appear black in poor light).

WOW!

Sometimes flocks of Blue Grosbeaks arriving in spring migration contain the similar but smaller Indigo Bunting. This is probably why one of the folk names for the Blue Grosbeak is Big Indigo.

LISTEN FOR: The song is a rich, burry warble with a few buzzy notes in the middle, delivered in a single unbroken phrase. The call note is a loud, emphatic *chink!*

REMEMBER: The male Blue Grosbeak looks like a chunky Indigo Bunting, but one look at its rusty shoulder patches and heavy bill, and you know you've got a Blue Grosbeak.

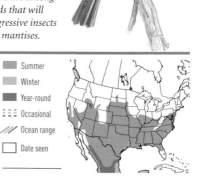

▶ *Blue Grosbeaks are among the few songbirds that will tackle large, aggressive insects such as praying mantises.*

FIND IT: Blue Grosbeaks often stay in thick cover in brambly thickets and along brushy field edges, but males will sing from treetop perches and fence and power lines in spring and summer. In migration, Blue Grosbeaks gather in loose feeding flocks.

Summer

Winter

Year-round

Occasional

Ocean range

Date seen

____

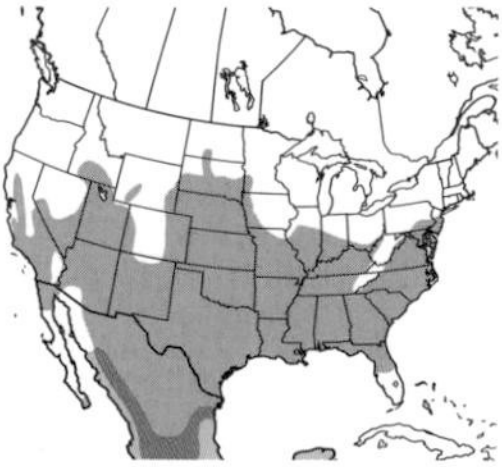

Location ______________

INDIGO BUNTING

Passerina cyanea Length: 5"

LOOK FOR: In fall the deep blue plumage of the adult male Indigo Bunting changes into the same drab brown plumage that the female wears all year. Young males may also be splotched with blue on the breast.

WOW!

A study with caged buntings inside a planetarium, recording the direction of their flight attempts as the star pattern above them was changed, proved that Indigo Buntings use the stars to navigate during migration.

LISTEN FOR: Choosing a treetop perch, male Indigo Buntings sing a loud, enthusiastic song of paired notes, often descending in tone: *fire! fire! where? where? there! there! put it out! put it out!* Call notes, frequently given, are a sharp *spick!* and a short, dry *bzzt!* that sounds like an electric shock.

REMEMBER: In very bright or very low light, the male Indigo Bunting appears all black, so relying solely on color to make an ID can be risky. Use size, shape, behavior, and sound to identify this species.

◀ *Drab brown female and young Indigo Buntings can be identified by the way they flick their tails out to the side and back.*

FIND IT: The Indigo Bunting can be heard singing along woodland edges and roadsides and in overgrown farm fields and other brushy habitat. Males are most conspicuous in spring and summer while females tend the nest.

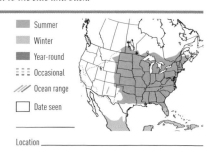

Summer

Winter

Year-round

Occasional

Ocean range

Date seen ____________

Location ____________

LAZULI BUNTING

Passerina amoena Length: 5½"

WOW!
Where Indigo and Lazuli Buntings nest in close proximity, they interbreed, and male Lazulis sometimes start to sing like male Indigos.

LOOK FOR: The adult male in breeding plumage is a bluebirdlike mix of blue head, back, and wings; rusty breast; and white belly. The main difference is the bold white wing bars on the Lazuli Bunting. Female Lazuli is plain brown overall with a buffy unstreaked breast and two buffy wing bars.

LISTEN FOR: Song is a goldfinchlike jumble of sweet whistled phrases: *twee-chiddledee-twee-twee two.* Call is a dry *spik.*

REMEMBER: If that blue bird singing from the top of a shrub has white wing bars, it's not a bluebird, it's a Lazuli Bunting.

◀ *As the human population has expanded and subdivisions have sprung up throughout the West, the Lazuli Bunting population has declined, along with those of many other species.*

FIND IT: Common in summer in open brushy areas, weedy hillsides, and in shrubs along streams. Males sing from a conspicuous perch during the breeding season. In migration, found in flocks in more open habitats, such as weedy fields and brushy meadows.

Summer
Winter
Year-round
Occasional
Ocean range
Date seen

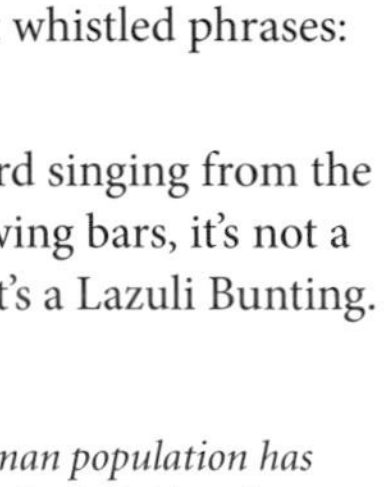

Location

PAINTED BUNTING

Passerina ciris Length: 5½"

LOOK FOR: Of all the birds in North America, the male Painted Bunting is the most vividly colored. Though not as colorful, the female is distinctive too, with a color combination no other bird has—lime green above and otherwise unmarked.

LISTEN FOR: The Painted Bunting's song is a long, sweet warbling phrase. It has the musical quality of the Indigo Bunting's song, but the notes are more slurred together. Call note is a sharp, metallic *vit!*

REMEMBER: Other small greenish birds (such as the vireos) do not have the Painted Bunting's large bill and overall unmarked plumage. Vireos have thinner bills and other obvious field marks (wing bars, eye lines, spectacles).

WOW!

Another name for the Painted Bunting is Nonpareil, which is French for "having no equal." On their tropical wintering grounds they are often illegally captured and kept as pet birds.

▶ *The female Painted Bunting cares for the young all by herself. Her mate may feed the fledglings when she starts a second brood.*

FIND IT: Despite the male's brilliant plumage, Painted Buntings can be hard to find. Their range is more limited than those of other buntings. They are shy birds that prefer thickets and brushy cover during the breeding season.

Summer
Winter
Year-round
Occasional
Ocean range
Date seen

____

Location ______________________________

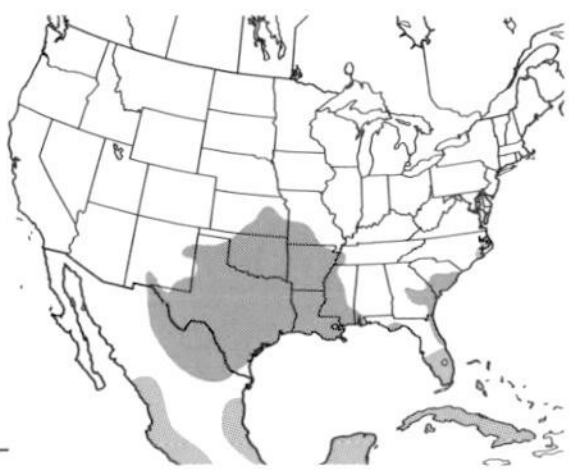

LARK BUNTING

Calamospiza melanocorys Length: 7"

LOOK FOR: Breeding-plumaged adult male is nearly all black with a large white wing patch and tail tips. Adult females and nonbreeding males are brown and streaky overall with a hint of the white wing patch and bold dark mustache lines on the sides of the throat. Heavy bill is blue-gray.

LISTEN FOR: Male Lark Buntings usually give their song during display flights near the breeding territory. Song is a melody of sweet whistles, trills, and buzzes. Many notes sound similar to song of Northern Cardinal. Call is a soft-noted *hyew.*

REMEMBER: Female and nonbreeding male Lark Buntings look like large streaky sparrows, but the super-heavy blue-gray bill is a great clue to their true identity.

WOW!

Because there are few song perches in their preferred habitat, male Lark Buntings perform their courtship and territorial songs while flying.

◀ *In flight, this mostly black bird has large white wing patches that really stand out.*

FIND IT: Common in summer on short-grass prairies and treeless open grasslands and prairies. During migration can be found in large flocks in any grassy, open habitat. Spends the winter in large flocks.

Summer

Winter

Year-round

Occasional

Ocean range

Date seen

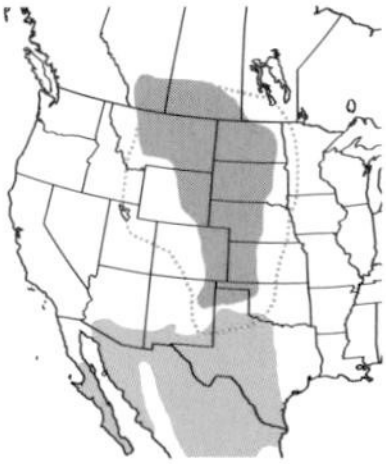

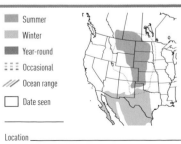

Location

Spiza americana Length: 6¼"

LOOK FOR: Looking like a miniature Meadowlark with its black V on a yellow breast, the breeding-plumaged male Dickcissel has a conical pale gray bill and a chestnut shoulder patch. Winter male is faded overall. Female looks like female House Sparrow but shows the rusty shoulder and a faint yellow wash on the breast.

WOW!

Dickcissels, like some other grassland-breeding birds, have a breeding range that expands and contracts from year to year, depending on the amount of local rainfall.

LISTEN FOR: Song is a series of percussive notes for which the bird is named: *dick-dick-dick-ciss-ciss-ell.* Call, usually given in flight, is a loud, burry *bzzrt!*

REMEMBER: The conical gray bill and rusty shoulder patch are field marks that set the Dickcissel apart from similar species in all seasons.

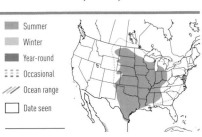

◀ *If you are in Dickcissel habitat in summer, you are likely to hear a male singing—they are very vocal birds.*

FIND IT: A common summer-breeding bird in grasslands, meadows, farm fields, and weedy pastures and fence-rows. Dickcissels are very vocal birds both during the breeding season and in migration, when their flight call can be heard even when the birds are flying too high to be seen.

Summer
Winter
Year-round
Occasional
Ocean range
Date seen

____

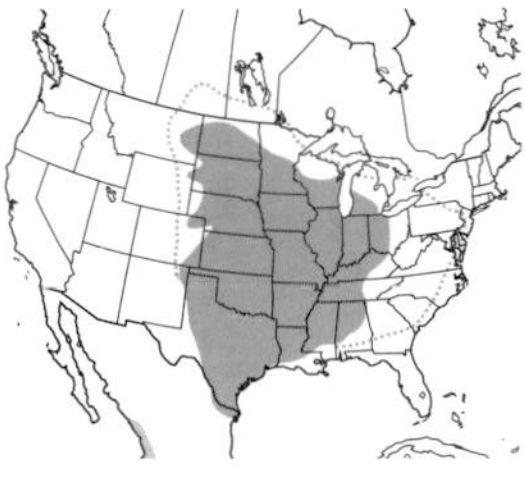

Location ____________________

EASTERN and WESTERN MEADOWLARKS

Sturnella magna, Sturnella neglecta Length: 9½"

LOOK FOR: The lemon yellow breast with a black V sets the Eastern Meadowlark apart from all other blackbirds (except for the look-alike Western Meadowlark). Males and females look alike.

LISTEN FOR: Eastern Meadowlarks sing a clear pattern of downward-slurring whistles: *see I SING clear! Or spring of THE year!* They utter several calls, including a harsh, electrical *jrrt!* and a sputtering rattle: *brtbrttttttt!*

REMEMBER: It might be impossible to separate Eastern and Western Meadowlarks visually. If you hear them sing, it's easier to make a positive identification. Westerns have a lower-pitched, less musical song that has a burbly ending. Eastern's song is high, slurred whistles.

◀ *A male Eastern Meadowlark sends his clear song over a rolling hay meadow.*

WOW!

Eastern Meadowlarks build an elaborate nest on the ground, woven out of grass. The nest often has a woven dome over it and an entrance on the side.

FIND IT: Common in spring and summer in grasslands, meadows, and prairies where males sing from a prominent perch. In winter, small flocks can be found in any open habitat, including cultivated fields and the grassy edges of airfields.

EASTERN MEADOWLARK

☐ Date seen ______________

Location ______________

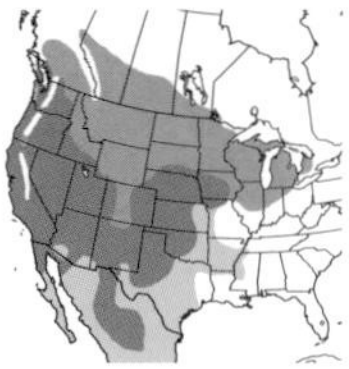

WESTERN MEADOWLARK

☐ Date seen ______________

____

WOW!
In parts of North America, the blackbird's **CONK-A-REE!** is one of the first sounds of spring's return. Males will begin singing on territory before the last snow has melted.

LOOK FOR: The adult male Red-winged Blackbird looks just like its name suggests—an all-black bird with obvious red (and yellow) shoulder patches. Males flash these patches as they fly and as they sing. Females are dark and streaky, with rusty backs and buffy eyebrows and throats.

LISTEN FOR: The blackbird's song is a harsh, rising *conk-a-ree!* Females give an explosive, sputtering call: *bee-bee-bee-prrrrrttt!* Flight call is *chack!* Alarm call of male is downslurred *tyeer!*

REMEMBER: The female Red-winged Blackbird looks confusingly similar to a streaky sparrow, so check the bill. The blackbird's dark bill is thinner and much longer than the typical sparrow bill.

▶ *When singing on territory, male Red-winged Blackbirds spread their wings to show off their red shoulder patches.*

FIND IT: Red-winged Blackbirds are found in spring and summer anywhere there is a bit of water and some tall grass, cattails, or other vegetation: marshes, lakes, ponds. Can be found in winter almost anywhere.

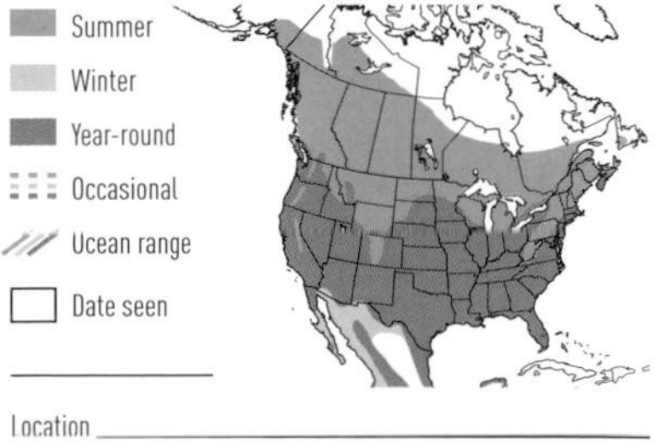

YELLOW-HEADED BLACKBIRD

Xanthocephalus xanthocephalus Length: 9½"

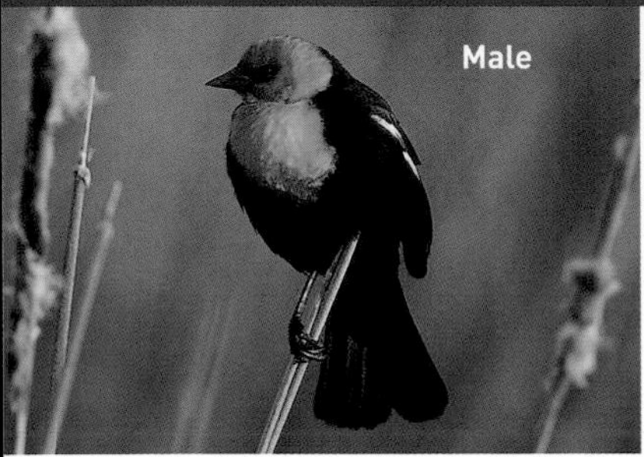

WOW!

More common in the Midwest and West, Yellow-headed Blackbirds do show up in winter blackbird flocks in the East. Sharp-eyed birders can pick them out of flocks of Red-winged Blackbirds and Common Grackles.

LOOK FOR: The male Yellow-headed Blackbird is a large, obvious black bird with a mustard yellow head and breast. His large white wing patches flash in flight. The female is drab brown with yellow on the face and chest.

LISTEN FOR: The male's song sounds like someone is throwing up. He points his bill upward and tilts his head to the side, showing off his colors. He does manage a few more musical notes now and then. Calls include a harsh, nasal *raar-raar-raar-raar!* and a loud, dry *chek!*

REMEMBER: This relative of the meadowlarks shares their yellow coloration but not their habitat. Blackbirds prefer marshy settings with cattails. Meadowlarks prefer grassy meadows.

▶ *The Yellow-headed Blackbird's love song ends in a bray that sounds like a dying donkey.*

FIND IT: Common in reedy marshes, the Yellow-headed Blackbird spends summers in noisy nesting colonies. Males perch on the tops of cattails and deliver their retching songs. Females dart to and from the nests.

Summer
Winter
Year-round
Occasional
Ocean range
☐ Date seen

____

Location ______________

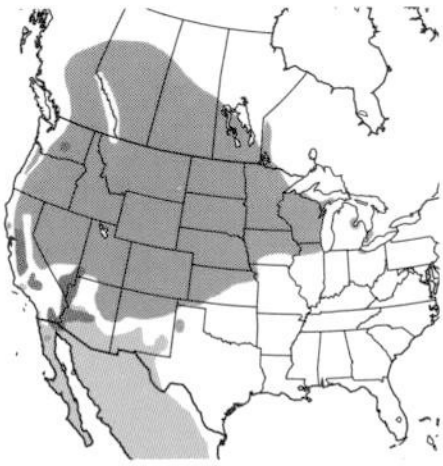

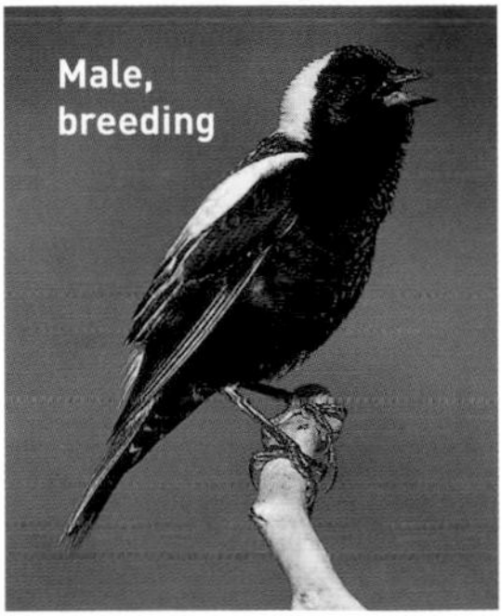

LOOK FOR: Male Bobolinks go through a dramatic plumage change from spring to fall. In summer, the male is a handsome blend of black, white, and butterscotch. In fall, he changes to buffy tans and browns, similar to what the female wears year-round.

LISTEN FOR: Male Bobolinks sing a burbling series of warbles, buzzes, and whistles in a long series that sounds like the start of a dial-up Internet connection or like R2D2 from *Star Wars.* Call note is a whistly, rising *wink?* Migrating Bobolinks often give this call in flight.

REMEMBER: Fall male and female Bobolinks may look like sparrows, but they are actually blackbirds. They are larger than our sparrows and have plainer-looking faces.

▼ *A male Bobolink in full song-flight over his nesting territory.*

WOW!

Other names for the Bobolink (a name derived from the song) are Meadow Wink (for its habitat and call note), Skunkhead Blackbird (plumage), and Butter-bird (for the yellow head patch).

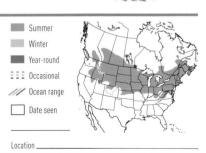

FIND IT: Locally common in grassy meadows and hay fields in spring and summer, the Bobolink is often located when the male does a song-flight over his territory. Females hide in the tall grass so are harder to see.

Summer
Winter
Year-round
Occasional
Ocean range
Date seen ______

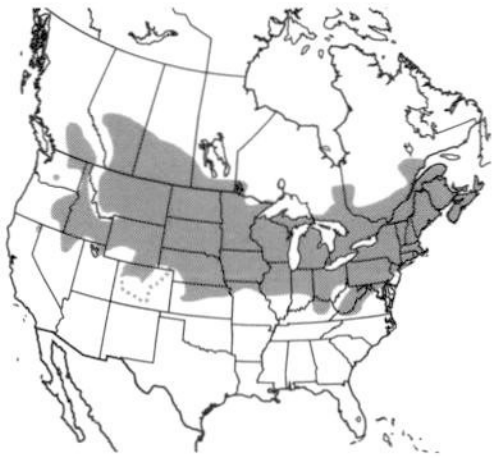

Location ______

LOOK FOR: In bright sunlight, the Common Grackle's feathers shine with tones of green, purple, gold, and blue. It's a dark bird that is larger than our other blackbirds (longer tail, heavier bill) but smaller than our other grackles. Females are slightly less colorful than males. All adults have pale yellow eyes (young birds have black eyes). In flight, the Common Grackle holds its tail in a V shape.

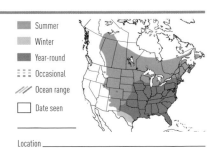

WOW!

Common Grackles will capture bats in the air and eat them. They also ambush House Sparrows in parks and near bird feeders, knock them on the head, and eat them. Eeewwww!

LISTEN FOR: Common Grackles make a variety of weird sounds. Their song is a series of squeaks, whistles, and grating scrapes: *krr-zheee! zhrrrt!* Call note, often given in flight, is a deep, sharp *cack!*

REMEMBER: Remember: Common Grackles fly straighter than blackbirds, which undulate like waves as they fly.

◄ *Acting more like a shrike than a blackbird, a Common Grackle flies off with a House Sparrow it took by surprise.*

FIND IT: The Common Grackle is found from suburban backyards and city parks to farm fields and wetlands. Nesting colonies are often built in large evergreen trees. Forages on the ground and visits bird feeders.

Summer

Winter

Year-round

Occasional

Ocean range

Date seen ________

Location ________

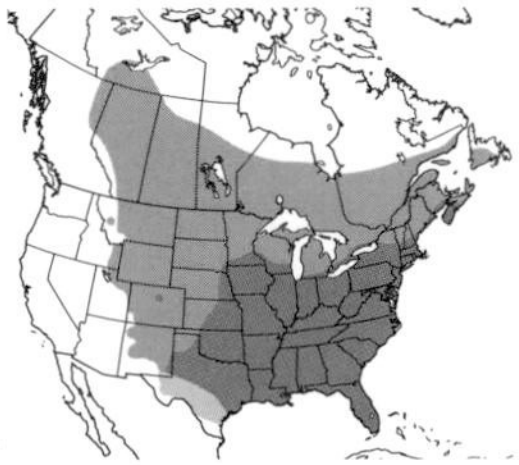

BOAT-TAILED GRACKLE

Quiscalus major   Length: 14½" (female) to 16½" (male)

LOOK FOR: This big dark bird is like a Common Grackle with a tail extension. Males are dark bluish green in bright sunlight. Females are dark brown overall with black wings and are slightly smaller than males.

LISTEN FOR: Voice is a harsh, repetitive series of buzzing trills, squeaks, and whistles: *krssshh-krssshh-krsssh, kweet-kweet, chaak-chaak-chaak!* Many of the Boat-tailed Grackle's sounds are very unbirdlike. Call note is a deep *chuck!*

REMEMBER: This species was formerly considered a single species with the Great-tailed Grackle, which lives in the Southwest. Great-taileds are slightly larger and not as closely associated with water as Boat-taileds.

WOW!

The Boat-tailed Grackle gets its name from the way it holds its tail in flight. The tail feathers form a V shape, like the keel of a boat.

▼ *Boat-tailed Grackles can be a nuisance, stealing pet food right off the porch. Females look like a different species from the males.*

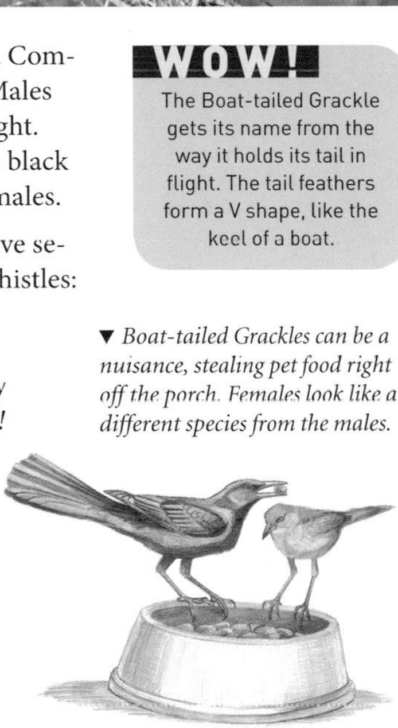

FIND IT: Very common and a permanent resident within its limited range, where it favors salt marshes and other open coastal areas. Boat-tailed Grackles nest in colonies near water and are seldom found far from water.

☐ Date seen ____________

Location ____________

Quiscalus mexicanus Length: 18"

Female

Male, courtship display

WOW!

Great-tailed Grackles roost communally, often in towns where parks and shopping centers provide both habitat and easy access to food. These roosts remain noisy all night long.

LOOK FOR: Adult male is a large, glossy black bird with a long black tail, for which it is named. Females and young birds are smaller and pale brown below, dark gray above, with a pale line over the eye. This grackle is noticeably larger, taller, and longer tailed than the Common Grackle.

LISTEN FOR: Song is a repeated series of loud shrieks, whistles, rattles, and harsh notes: *weet-weet-weet. Chack-chack-chack. Boit-boit-boit. Annnkannkannk.*

REMEMBER: The similar Boat-tailed Grackle of the Southeast is never found far from the coast.

▶ *Male Great-tailed Grackles fluff up their feathers, quiver their wings, point their bills skyward, and utter their weird songs, all to charm the grackle gals.*

FIND IT: Common and expanding its range, the Great-tailed Grackle is impossible to miss wherever it occurs because of its large size and loud voice. Nearly always found in flocks. Forages in farmland, open groves, and feedlots. Roosts and nests in thick cover often near water.

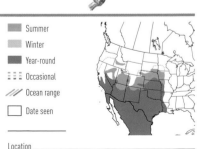

Summer

Winter

Year-round

Occasional

Ocean range

☐ Date seen

____

Location ______________

LOOK FOR: Adult male is all black with a pale eye. Plumage can appear glossy purple and green in bright sunlight. Tail is square ended and much shorter than those of grackles. Female Brewer's is plain grayish brown overall with a dark eye (female Rusty Blackbird has a pale eye).

LISTEN FOR: Song is shrill and unmusical: *ksh-ree!* Call is a loud *chak!*

REMEMBER: Telling breeding-season male Rusty and Brewer's Blackbirds apart can be tricky. Adult male Brewer's never takes on any rusty color and has a slightly thicker, shorter bill than the Rusty.

WOW!

This species was discovered by John James Audubon on his final expedition in 1843 and named for his friend Dr. Thomas Brewer, a Boston ornithologist, who probably never saw a live Brewer's Blackbird.

◀ *The Brewer's Blackbird has adjusted very well to human-altered habitats. "Would you like fries with that?"*

FIND IT: Very common in a variety of habitats in the West, from parking lots to farm fields and mountain meadow edges. Often seen on ground, walking, jerking its head forward with each step, much like a chicken. Joins huge mixed-species blackbird flocks in winter.

Summer
Winter
Year-round
Occasional
Ocean range
Date seen ______

Location ______

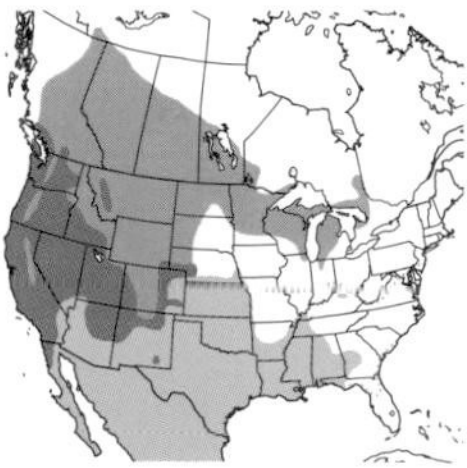

BROWN-HEADED COWBIRD

Molothrus ater Length: 7½"

LOOK FOR: This smallish blackbird has a short, stout bill and a short tail. The bill shape is a good field mark to separate this species from other blackbirds. The name *cowbird* comes from its habit of following and foraging around herds of cattle.

WOW!

Cowbirds are notorious for laying their eggs in the nests of other birds. The nestling cowbirds often outcompete smaller nestmates. This has caused a decline in many songbird populations.

LISTEN FOR: The song is a weird mix of low gurgles and high, squeaky whistles. Also gives a long sputtery trill: *pt-pt-pprrrrrrrtttt!* Flight call is a high, thin whistle: *tsee-tseeeet!*

REMEMBER: The European Starling is often seen in the same habitat as the cowbird but differs by having a long, thin bill and a very short tail.

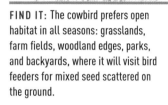

◀ *A female Brown-headed Cowbird removes a Wood Thrush egg and will lay one of her own. She may lay 30 eggs in a single season.*

FIND IT: The cowbird prefers open habitat in all seasons: grasslands, farm fields, woodland edges, parks, and backyards, where it will visit bird feeders for mixed seed scattered on the ground.

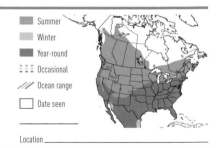

Summer
Winter
Year-round
Occasional
Ocean range
Date seen

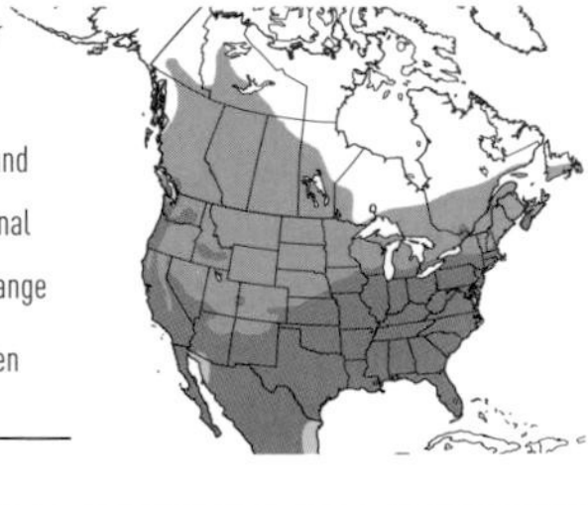

____

Location __________

WOW!

Orchard Orioles are not as territorial as some other songbirds. They commonly nest in a tree containing the nests of one or more other pairs of Orchard Orioles.

LOOK FOR: The Orchard Oriole is small enough to be mistaken for a warbler, but one look at the long, fine-pointed bill should convince you that this is no warbler. The male Orchard is a deep rusty orange and black. Females are plain yellow-green overall.

LISTEN FOR: Male Orchard Orioles sing a rapid, variable song that is sweetly musical with a few buzzy notes: *look here! up here! see me? how's it going! chh-chh! hey you!* Call note is a soft *chuck!* or *twee-ohh!*

REMEMBER: Female Orchard and Baltimore Orioles can be hard to tell apart. Female Orchards are smaller and greener overall; female Baltimores are washed with orange on chest and belly.

▶ *A female Orchard Oriole tends her nestlings in an apple tree.*

FIND IT: Common in summer in young woods and woodland-edge habitat, Orchard Orioles prefer open, brushy settings and avoid deep woods. Males are avid treetop singers.

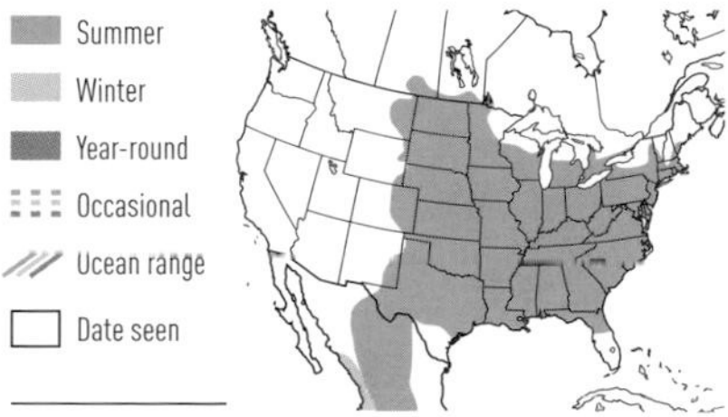

Date seen ______

Location ______

Icterus galbula Length: 8¾"

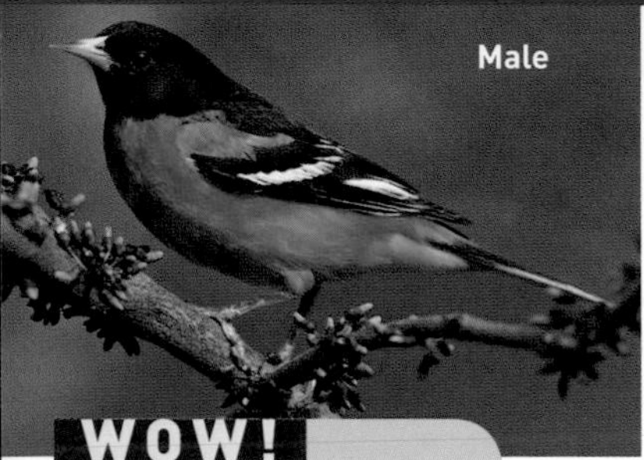

WOW!

The Baltimore Oriole was named for the royal colors of Lord Baltimore. Today we have a beloved bird and a major-league baseball team that proudly bear this name.

LOOK FOR: The classic orange and black colors of the male Baltimore Oriole are familiar even to nonbirders. No other eastern bird species has the male Baltimore's combination of size and coloration.

LISTEN FOR: The Baltimore Oriole's song is a series of clear, slurred whistles. It varies in pattern: *hey! hey you! see me up here sing-ing!* Call is a scolding chatter: *chrt-trr-rrrrrr!*

REMEMBER: The Baltimore Oriole was once lumped together with the Bullock's Oriole (a western species) as the single-species Northern Oriole. These birds overlap in range and do interbreed, but the males are easy to separate visually (the male Bullock's Oriole has an orange face).

▶ *Baltimore Orioles will visit feeders for fruit, such as orange halves.*

FIND IT: The Baltimore Oriole prefers open woodland settings with tall trees for nesting, such as in city parks and farmland groves. Its baglike nest is often built near the end of a horizontal branch overhanging a road or water.

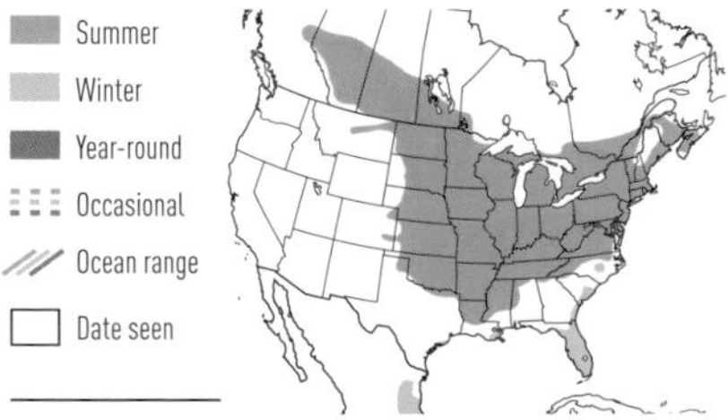

Location ______________________

BULLOCK'S ORIOLE

Icterus bullockii Length: 8¼–8½"

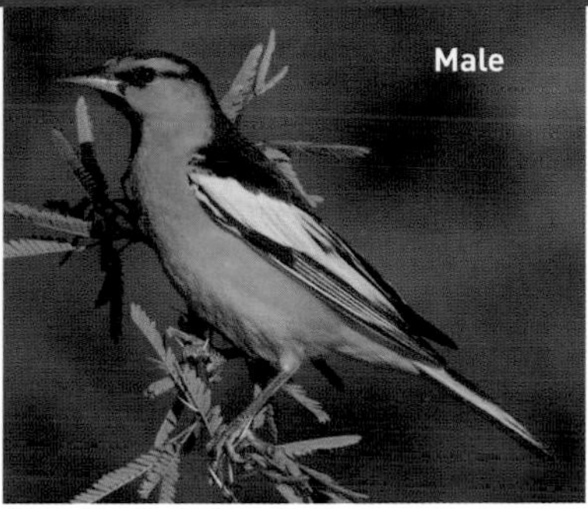

WOW!

A group of ornithologists lumped together the Bullock's Oriole and the Baltimore Oriole into a single species: Northern Oriole. Then some other ornithologists split them into two species again. Stay tuned!

LOOK FOR: Adult male is bright orange below with an orange face, black throat and crown, and black eye line. Huge white wing patch (on the coverts) stands out on black wings and is obvious in flight, as is black-tipped tail of male. Female is pale orange on the head and gray below, with a gray back and subtle white wing bars.

LISTEN FOR: Song is a series of paired notes, some musical, some harsh: *cha-chacha-toowee-tricka-trickatricka-reeet!* Call is a long scolding chatter: *ch-ch-ch-ch-ch-ch.*

REMEMBER: The male Bullock's Oriole has a black crown and eye line. The similar Hooded Oriole has an orange crown and nape.

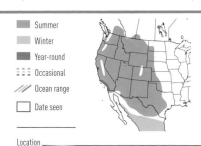

▶ *Once they identify it as a food source, Bullock's Orioles will visit a hummingbird feeder for nectar.*

FIND IT: Common in summer in deciduous woods, especially in cottonwoods along streams, where Bullock's Orioles sometimes form loose nesting colonies. Males are very vocal on breeding territory.

Summer

Winter

Year-round

Occasional

Ocean range

Date seen ______

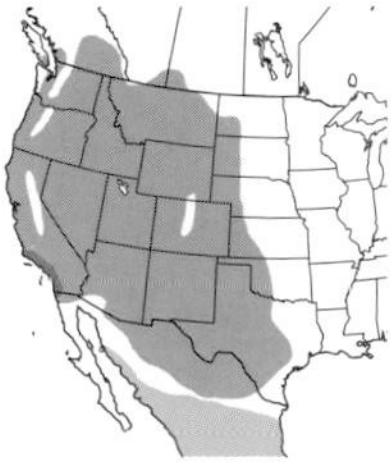

Location ______

HOODED ORIOLE

Icterus cucullatus Length: 7½–8"

LOOK FOR: Adult male is orange and black with an orange crown and black face and throat. Bill is thinner and down-curved. White coverts on black wings are less extensive than those of Bullock's Oriole. Female Hooded Oriole is olive above, yellow below.

WOW!

Some Hooded Orioles spend the winter in the U.S. in places where nectar feeders guarantee a steady food supply.

LISTEN FOR: Song is a jumble of whistles, squeaks, and chattering notes. Calls include a rising *wheet!* and a short, soft chatter.

REMEMBER: The "hood" on the Hooded Oriole is golden orange, not black as in many of our other orioles.

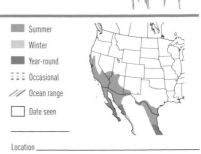

► *Hooded Orioles nearly always choose to nest in palm trees, particularly fan palms.*

FIND IT: Uncommon in the desert Southwest in summer in open woods, palm and streamside groves, parks, and wooded suburban neighborhoods. Usually associated with palm trees. A common visitor to backyard feeders for nectar and fruit.

Summer

Winter

Year-round

Occasional

Ocean range

Date seen

Location

GRAY-CROWNED ROSY-FINCH

Leucosticte tephrocotis Length: 6–8"

LOOK FOR: Adult has gray head, black forehead and throat, and varying amounts of rose in the body and wing feathers. Breeding male has a blacker face, a black bill, and bright rosy patches on wings and flanks.

LISTEN FOR: For such a beautiful finch, the Gray-crowned Rosy-finch's song is nothing special: loud down-slurred *tyeew-tyeew* notes in a series, similar to the call of the House Sparrow. Flight call is a softer *tchew!* Flocks are very vocal.

REMEMBER: This species is highly variable geographically and often flocks with other rosy-finches in winter, so it pays to look at each bird individually.

WOW!

Some high-altitude feeding stations, such as the one at Sandia Crest, near Albuquerque, New Mexico, attract all three species of rosy-finch in winter: Gray-crowned, Black, and Brown-capped!

▶ *Most birders get their life looks at Gray-crowned Rosy-finches that are visiting a feeding station.*

FIND IT: This is our most widespread rosy-finch, but its habitat preferences make it difficult to find: high mountain meadows, tundra, rocky summits, and snowfields. Usually in flocks. Often visits feeders at ski resorts and in mountain towns.

Summer

Winter

Year-round

Occasional

Ocean range

Date seen

____

Location ______________

WHITE-WINGED CROSSBILL

Loxia leucoptera Length: 6½"

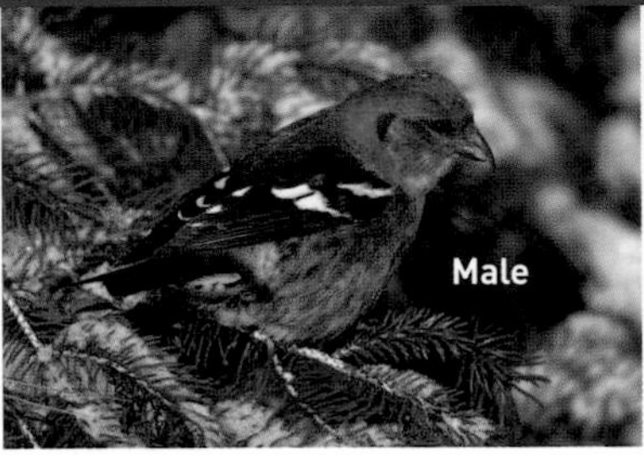

LOOK FOR: Adult male is pinkish red overall with bold white wing bars on black wings. Females and young birds are brownish and very streaky overall; they also have white wing bars.

LISTEN FOR: Song is a long unmusical rattle that slows near the end. More commonly heard is the call, given in short phrases: *chi-dit, chi-dit, chi-dit!*

REMEMBER: White-winged Crossbills are named for their best field mark (white wing bars). Red Crossbills lack any white in the wings and are more brick red in color.

WOW!

White-winged Crossbills wander constantly in search of abundant spruce cones. When they find them, they will often stop to nest, even in midwinter!

▶ *Flocks of foraging White-winged Crossbills may contain hundreds of individuals. Sometimes they are joined by Red Crossbills and other finch species.*

FIND IT: A vagabond species of the far North that is highly reliant on the variable crop of spruce cones. Always found in flocks in spruce, hemlock, and fir forests.

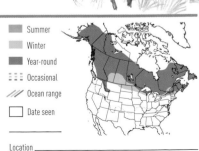

Summer

Winter

Year-round

Occasional

Ocean range

Date seen ________

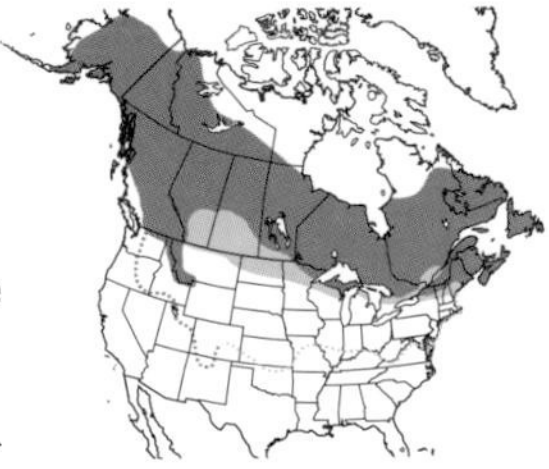

Location ________

LOOK FOR: This bird's name is very descriptive: The male is a red bird with a weird crossed bill that is perfectly designed to pry the seeds out of evergreen cones. Females are dull yellow-green and have dark wings like the males. The Red Crossbill does not have wing bars, unlike its smaller-billed cousin the White-winged Crossbill.

> **WOW!**
> There are at least nine distinct types of Red Crossbill in North America. Some have very large bills and feed on large pinecones. Others have smaller bills and eat seeds of smaller cones.

LISTEN FOR: A short series of thin, buzzy chips passes for the song of the Red Crossbill: *twit-twit-twit-twit, jwee-zit.* Call note is a loud, hard *klip-klip!*

REMEMBER: The crossed bill can be difficult to see from a distance. But the way crossbills feed—often hanging upside down from a pinecone, digging with the bill to pry seeds loose—is distinctive.

▶ *A Red Crossbill uses its scissorlike mandibles to extract spruce seeds.*

FIND IT: Red Crossbills are rarely found far from conifers (pines, hemlocks, spruces, firs) and are best located by the sounds made by feeding flocks.

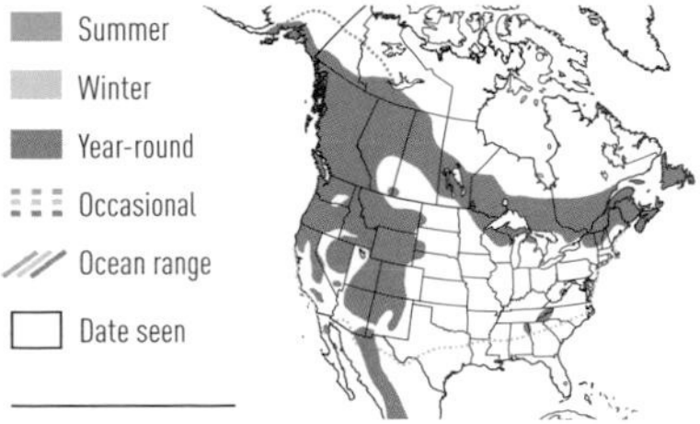

Date seen ______________

Location ______________

COMMON REDPOLL

Carduelis flammea Length: 5"

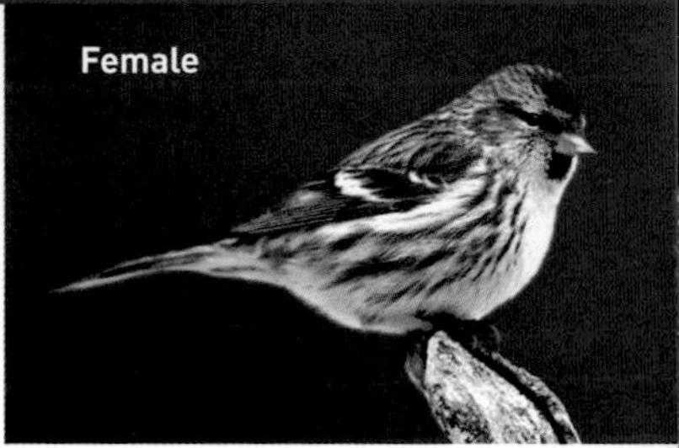

LOOK FOR: The smallest of our red finches, the Common Redpoll looks like a Pine Siskin with a red cap on its head. Streaky on backs and sides, with a tiny yellow bill surrounded by a black chin, the Common Redpoll is a very active, vocal bird. The male may have a completely pink-washed chest, while the female has a white chest.

LISTEN FOR: Song is a long series of twitters and rising buzzy trills: *chit-chit-chit-chewee, tu-tu-tu-tseeet, chit-chitchit-zeeeet!* Common call is a chattering *ch-ch-ch-chweee!*, rising in tone on the last, longer note.

▼ *Common Redpolls forage hastily, filling special pouches in their upper gut. Then they regurgitate the seeds, shell them, and eat them.*

REMEMBER: The Common Redpoll's small size and red cap make it easy to separate from other red or streaky finches.

WOW!

The polar bears of the bird world, redpolls can survive colder temperatures better than any other songbird. Extra food, stored in their crops, is digested during the night, keeping the redpoll warm.

FIND IT: In winters when food is scarce in the North, redpoll flocks, sometimes mixed with goldfinches and siskins, forage actively in weedy fields, trees, and shrubs. They will visit bird feeders for thistle seed and sunflower seed bits.

☐ Date seen

____

Location ______________

Carpodacus mexicanus Length: 5¾"

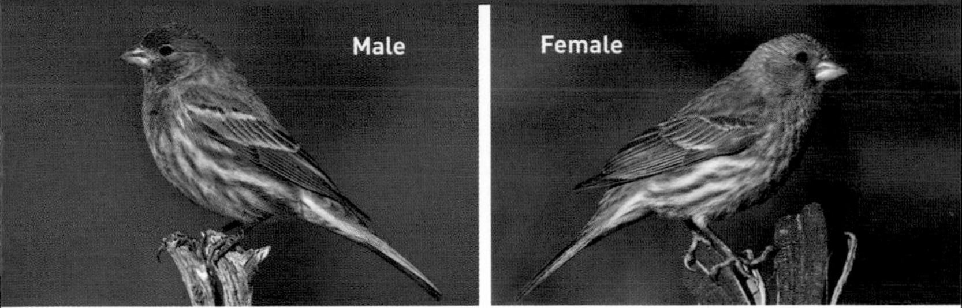

LOOK FOR: The male House Finch is washed with brick red on head and breast and streaked with brown everywhere else, especially on the flanks (sides). Females and young birds are brownish gray overall and covered with blurry streaks.

WOW!

In 1940, a small number of House Finches being sold illegally as caged birds were released near New York City. All the House Finches east of the Great Plains are descended from them.

LISTEN FOR: The song is a rich series of whistled phrases ending with a few buzzy notes. Also utters a variety of call notes, from sparrowlike chirps to a rising *kweet!*

REMEMBER: Male House Finches can vary in color from brick red to orange, but they always have blurry dark streaks on their bellies. They look dingy compared to the cleaner, brighter male Purple Finch (which has an unstreaked white belly).

► *A pair of House Finches enjoy the seeds they find in birch cones.*

FIND IT: A native of the West, with the help of humans the House Finch has colonized the East. They are adaptable to nearly any habitat except deep woods and open grasslands, and most common near human-affected areas.

Summer
Winter
Year-round
Occasional
Ocean range
Date seen

Location ________________

LOOK FOR: The male Purple Finch is washed with a raspberry red (not purple), as if he'd been dipped upside down in raspberry juice. Unstreaked white flanks, belly, and wings help separate him from the similar male House Finch. The female is covered in short dark brown streaks.

WOW!

Huge influxes of Purple Finches move south in fall and winter when the natural food crop (seeds of pines, spruces, and other trees) is poor in the North.

LISTEN FOR: The bright, cheery song is a fast series of hoarse whistles: *treedle-wheedle-treedle-turtle-wheedle-breer!*

REMEMBER: House Finches just don't look as clean as Purple Finches do. The Purple Finch also has a larger bill, which gives it a larger-looking head. Check out the female Purple Finch's white eye line and dark bandit mask. By contrast, the female House Finch looks plain faced.

◀ *A male Purple Finch pulls out all the stops, singing for a female. At the height of his display, he may almost tip over backward.*

FIND IT: Purple Finches are conspicuous birds, perching in treetops and calling and singing frequently. Flying birds can be easily located by the flight call (*pit-pit!*).

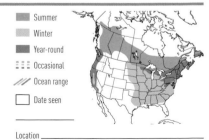

Summer

Winter

Year-round

Occasional

Ocean range

☐ Date seen

____

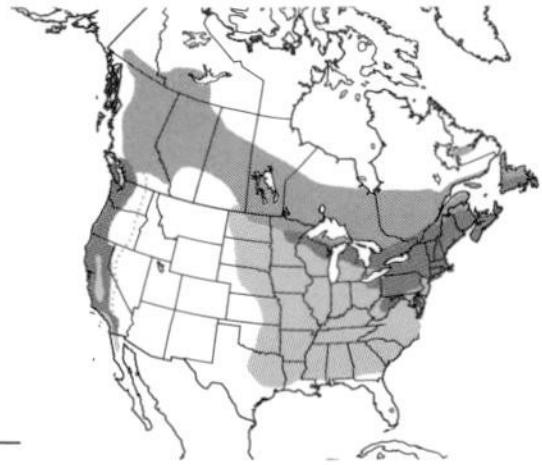

Location ______________

LOOK FOR: Adult male has a crown that is brighter red than the rest of its head and neck, giving it a peaked-headed appearance. Pinkish red on breast is paler than extensive red of male Purple Finch. Female Cassin's is finely streaked with brown overall, especially on the breast and belly. Cassin's Finch has a longer, more pointed bill than either Purple or House Finches.

LISTEN FOR: Song is a rich musical warble, softer in tone than song of Purple Finch. Call is a rising *giddy-up!*

REMEMBER: Red finches in the western mountains should be examined closely. The longer, more pointed bill and the red cap of the male Cassin's Finch set it apart from both Purple and House Finches.

▲ *After nesting, Cassin's Finches like to roam around in flocks, often mixing with crossbills and Evening Grosbeaks.*

WOW!

Cassin's Finches (and the uncommon Cassin's Sparrow) were named for John Cassin, a nineteenth-century ornithologist from Philadelphia, a place where Cassin's Finches never occur!

FIND IT: A resident of mountain coniferous forest of the West and usually found in flocks, except during actual nesting. Found at high altitudes in summer, and some flocks move to lower elevations in winter.

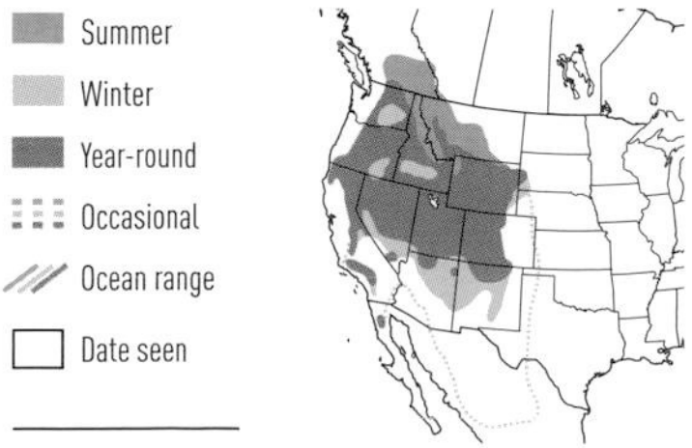

Location ____________

Male (left), female (right)

LOOK FOR: The striking male Evening Grosbeak is a large mustard yellow finch with a big pale bill and a gold eyebrow. Huge white patches cover the bird's back and flash in flight on the black wings. Females are mostly gray with light yellow patches and less white in the wings. Coloration does not change seasonally.

WOW!

The Evening Grosbeak was named for the mistaken impression that it sang only at dusk.

LISTEN FOR: Loud and social, Evening Grosbeaks announce their presence with constant calling. Common call is a loud, descending *pyeer!* Also utters a raspy, tuneless whistle: *pirrrt!*

◀ *The Evening Grosbeak can exert as much as 125 pounds of pressure with its bill and can crack cherry pits.*

REMEMBER: Like other winter finches, the Evening Grosbeak has a swooping flight style. But only the Evening Grosbeak flashes large white wing patches in flight.

FIND IT: Listen for the Evening Grosbeak's call note to find this unpredictable wanderer. Outside of nesting season, flocks move to where the food is—maples and box elders for their seeds and bird feeders for sunflower seed.

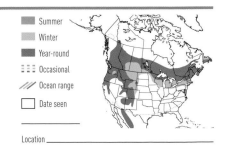

Summer
Winter
Year-round
Occasional
Ocean range
Date seen

____

Location ______________________________

AMERICAN GOLDFINCH

Carduelis tristis Length: 5"

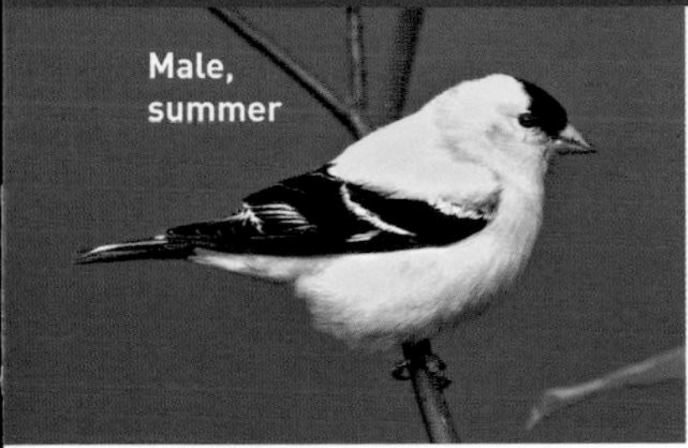

WOW!
The American Goldfinch is the only songbird that feeds its young a diet of seeds. Brown-headed Cowbirds hatched from eggs laid in goldfinch nests cannot survive the all-seed diet.

LOOK FOR: The American Goldfinch in spring and summer is familiar even to nonbirders. Bill color is yellow in spring and summer, dark gray in fall and winter. Winter-plumaged males resemble females, fooling many feeder watchers into thinking their goldfinches have left.

▲ *American Goldfinches eat thistle seeds and line their nests with thistle fluff.*

LISTEN FOR: The goldfinch's song is a variable mix of rich warbles and twitters, often given from a treetop perch. Twittering *potato-chip, potato-chip* call is commonly given in flight.

REMEMBER: Dull-colored winter goldfinches are often overlooked as they forage with other species. Similar in appearance to yellow-colored warblers, the American Goldfinch has a much heavier bill.

FIND IT: American Goldfinches forage in noisy, twittering flocks. They are common continent-wide in brushy habitat, along woodland edges, and in overgrown meadows. They visit feeders for sunflower and thistle seeds.

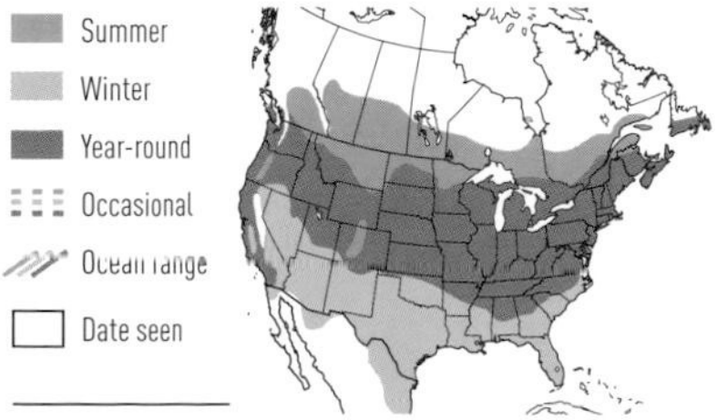

Date seen ____________

Location ____________

LESSER GOLDFINCH

Carduelis psaltria Length: 4½"

LOOK FOR: The Lesser Goldfinch is so named because it is our smallest goldfinch. Adult male is dark olive above with a black cap and yellow below. Female is pale olive above and yellow below, very similar to female American Goldfinch.

LISTEN FOR: Song is a long rapid series of sweet notes and whistles, often repeated: *sweee-cheecheechee-tvee-tvee-tvee-too-tootoo-sweee!* Often imitates phrases of other birds' songs.

REMEMBER: Breeding-plumaged male American Goldfinch has a canary yellow back and white rump. Lesser Goldfinch has a dark back and rump.

WOW!

The back color of adult male Lesser Goldfinches varies geographically. Males in the southern Rocky Mountains have black backs, while most others have olive green backs.

▶ *The diet of the Lesser Goldfinch is mostly plant seeds. They seem to prefer thistle and sunflower seeds above all others.*

FIND IT: Common, but sometimes overlooked, in open brushy habitat, weedy fields, streamsides, parks, and gardens. Usually found in flocks actively foraging for seeds, uttering soft twittering vocalizations.

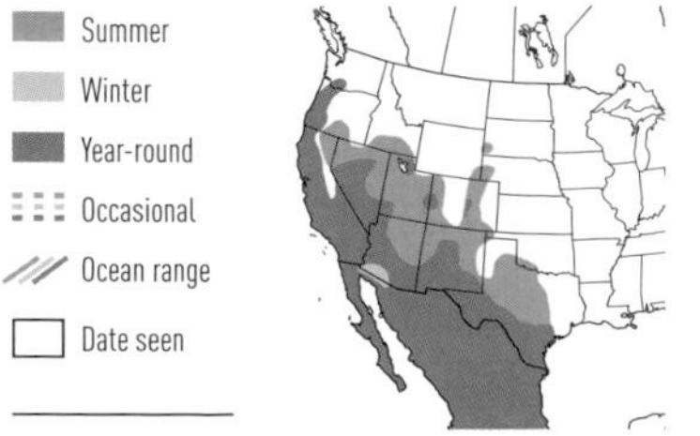

Location ____________

LOOK FOR: This handsome bird is mostly grayish. Adult male has a black face, yellow breast, and large yellow patches in the wings and on the rump. Female has an all-gray head and less yellow elsewhere. In flight, both sexes show white underwing linings and a yellow rump.

WOW!

This uncommon species can often be found near sources of water in remote arid areas.

LISTEN FOR: Lawrence's Goldfinch has one of the more interesting finch songs, a long melody of musical phrases, most of which are imitations of other birds' songs. Call is a slightly harsh, rising *too vreee!*

REMEMBER: Flocks of goldfinches in the desert Southwest may include this species. It's worth carefully sorting through these flocks.

► *Lawrence's Goldfinches are quite fond of salt and will visit a reliable source of this mineral.*

FIND IT: Uncommon in mixed oak-pine woods, chaparral, weedy areas usually near water. Found almost exclusively in California, except in winter, when flocks may range farther to the east, though never leaving the southwestern desert.

Date seen ______

Location ______

LOOK FOR: This streaky brown finch has a narrow, fine-tipped bill and shows yellow flashes in wings and tail in flight. They are often overlooked when mixed in with winter flocks of American Goldfinches, which are drab yellow (but not streaky) in nonbreeding plumage. The siskin's heavy streaking also makes it resemble a female House Finch, but note the siskin's smaller, finer bill.

LISTEN FOR: Pine Siskins are very vocal birds and commonly utter a harsh, rising call: *zzzrrreeeee?* Also a variety of thin, high twitters, often in paired notes: *twee-twee, jee-jee, twee-twee!*

REMEMBER: Pine Siskins have been called "goldfinches in camouflage." They act, fly, and forage like American Goldfinches.

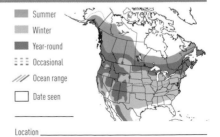

◀ *A male Pine Siskin feeds his mate on the nest in a balsam fir.*

WOW!

In the spring following "invasion" winters, when thousands of Pine Siskins come south for food, some siskins stay behind to nest, beyond the boundaries of the normal breeding range.

FIND IT: Pine Siskins often associate with flocks of goldfinches foraging in weedy fields and visiting bird feeders. Flocks of Pine Siskins call frequently, so learning to recognize their vocalizations is helpful.

Summer

Winter

Year-round

Occasional

Ocean range

Date seen ________

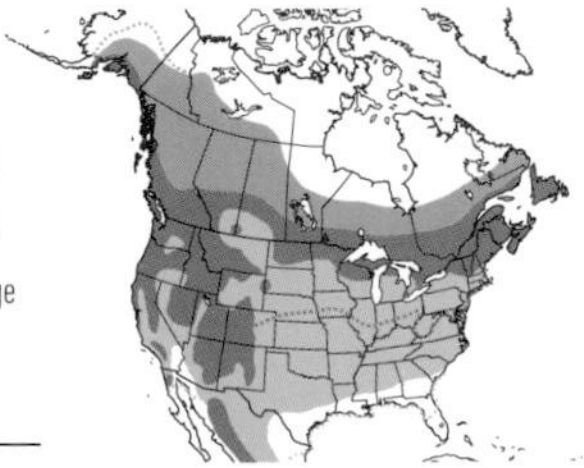

Location ________

HOUSE SPARROW

Passer domesticus Length: 6¼"

WOW!

House Sparrows were imported from Europe and introduced in New York in 1850 to control insects on city streets. By the early 1900s, they had spread across the continent to California.

LOOK FOR: This chunky, big-headed sparrow is a common sight in small flocks in cities, backyards, and farmyards. In winter, the male's colors are less distinct than in spring and summer. Females are drab gray-brown year-round. Flight is direct and swift with rapid wingbeats.

LISTEN FOR: House Sparrows do not really sing a song. Instead, they constantly utter a series of husky calls: *cheerp! cha-deep!* They also give harsh rattles and short whistles near the nest.

REMEMBER: The House Sparrow thrives in cities where other sparrows cannot survive. It is never found far from human habitation, so a sparrow you see in natural, unaltered habitat is almost certainly not a House Sparrow.

▶ *House Sparrow nests are very messy, with lots of long grass stems and even bits of trash mixed in.*

FIND IT: Common in any human-affected landscape, House Sparrows nest in cavities (woodpecker holes, nest boxes) and often compete with native songbirds. House Sparrows visit bird feeders, eating almost anything that's served.

Summer
Winter
Year-round
Occasional
Ocean range
☐ Date seen

____

Location ___________

RESOURCES

NATIONAL ORGANIZATIONS FOR BIRD WATCHERS

American Bird Conservancy
PO Box 249
The Plains, VA 20198-2237
540-253-5780
www.abcbirds.org

American Birding Association
93 Clinton Street
Delaware City, DE 19706
800-850-2473
www.aba.org

Cornell Laboratory of Ornithology
159 Sapsucker Woods Road
Ithaca, NY 14850
800-843-2473
www.birds.cornell.edu

National Audubon Society
225 Varick Street
New York, NY 10014
212-979-3000
www.audubon.org

National Wildlife Federation
11100 Wildlife Center Drive
Reston, VA 20190
800-822-9919
www.nwf.org

The Nature Conservancy
4245 N. Fairfax Drive
Suite 100
Arlington, VA 22203-1606
800-628-6860
www.nature.org

FIELD GUIDES TO BIRDS

Brinkley, Edward S. *National Wildlife Federation Field Guide to Birds of North America*. New York: Sterling Publishing, 2007.

Crossley, Richard. *The Crossley ID Guide to Eastern Birds*. Princeton, NJ: Princeton University Press, 2011.

Dunn, Jon L., and Jonathan Alderfer. *National Geographic Field Guide to the Birds of North America*. 6th ed. Washington, DC: National Geographic, 2011.

Floyd, Ted. *Smithsonian Field Guide to the Birds of North America*. New York: HarperCollins, 2008.

Kaufman, Kenn. *Kaufman Field Guide to Birds of North America*. Boston: Houghton Mifflin, 2005.

Peterson, Roger Tory. *Peterson Field Guide to Birds of North America*. Boston: Houghton Mifflin, 2008.

Robbins, Chandler S., et al. *Birds of North America: A Guide to Field Identification*. Revised ed. New York: St. Martin's Press, 1983.

Sibley, David Allen. *The Sibley Guide to Birds*. New York: Knopf, 2000.

Sterry, Paul, and Brian E. Small. *Birds of Eastern North America: A Photographic Guide/Birds of Western North America: A Photographic Guide*. Princeton, NJ: Princeton University Press, 2009.

Stokes, Donald, and Lillian Stokes. *The New Stokes Field Guide to Birds: Eastern Region/Western Region*. Boston: Little, Brown and Co., 2013.

Thompson, Bill, III. *The Young Birder's Guide to Birds of North America*. Boston: Houghton Mifflin Harcourt, 2012.

Vuilleumier, François, ed. *Birds of North America, Eastern Region/Western Region* (American Museum of Natural History). New York: Dorling Kindersley, 2009.

AUDIO GUIDES TO BIRDS

Elliott, Lang. *Know Your Bird Sounds*. Vols. 1 and 2. Mechanicsburg, PA: Stackpole Books, 2004.

Elliott, Lang, with Donald and Lillian Stokes. *The Stokes Field Guide to Bird Songs: Eastern Region*. Boston: Little, Brown, 2010.

Peterson, Roger Tory, ed. *A Field Guide to Bird Songs: Eastern and Central North America*. Revised ed. Boston: Houghton Mifflin, 2002.

Walton, Richard K., and Robert W. Lawson. *Birding by Ear: Eastern/Central*. Boston: Houghton Mifflin, 2002.

———. *More Birding by Ear*. Boston: Houghton Mifflin, 2000.

FIELD GUIDE AND BIRD SONG APPS FOR BIRD WATCHERS

Audubon Birds: A Field Guide to North American Birds (Green Mountain Digital). www.audubonguides.com.

BirdJam HeadsUp Warblers/HeadsUp Sparrows (MightyJams LLC). www.birdjam.com.

BirdsEye (Birds in the Hand). www.birdseyebirding.com.

BirdTunes (BirdTunes Encyclopedia of Bird Songs). www.birdtunesapp.com.

eBird. www.ebird.org.

iBird Explorer (Mitch Waite Group). www.ibird.com.

National Geographic's Handheld Birds (National Geographic Society). www.nationalgeographic.com/mobile/apps/handheld-birds/.

Peterson Birds of North America (Appweavers, Inc.). www.petersonguides.com.

Sibley eGuide to Birds of North America (mydigitalearth.com). www.sibleyguides.com.

PERIODICALS FOR BIRD WATCHERS

Watching Backyard Birds Newsletter, PO Box 110, Marietta, OH 45750, 800-879-2473. www.watchingbackyardbirds.com.

Bird Watcher's Digest, PO Box 110, Marietta, OH 45750, 800-879-2473. www.birdwatchersdigest.com.

BirdWatching, Madavor Media, 25 Braintree Hill Office Park, Suite 404, Braintree, MA 02184. www.birdwatchingdaily.com.

Living Bird, Cornell Laboratory of Ornithology, 159 Sapsucker Woods Road, Ithaca, NY 14850, 800-843-2473. www.allaboutbirds.org.

GLOSSARY

Bib: the area below a bird's bill covering the throat and upper breast. Refers to same area covered by a bib worn by a human.

Binocs: abbreviation used by birders for binoculars.

Breeding plumage: the set of feathers worn during the breeding season (spring and summer). This is often the most colorful plumage, especially among male songbirds. See *nonbreeding plumage.*

Brood parasitism: a behavioral habit characterized by birds laying their eggs in the nests of other birds.

Cache: a place where food is stored or hidden for later consumption.

Call note: brief, relatively simple sound uttered by birds in various social contexts (for example, location calls, food calls).

Cavity nester: a bird that nests inside an enclosed area, such as a hollow tree, an old woodpecker hole, or a nest box.

Central Flyway: the migration route used by birds through the central portion of North America from central Canada, across the Great Plains, and southward to the Gulf of Mexico.

Checklist: a list of bird species compiled from records in a specific geographic area.

Coverts: the small contour feathers on the upper part of a bird's wing that overlap the flight feathers.

Crest: a tuft of long feathers on a bird's head that may be held erect.

Crissum: the feathers covering the undertail of a bird, as in the rust-colored crissum of the Gray Catbird.

Crown: the top of a bird's head. (See "Parts of a Bird," page 20.)

Dabbling: refers to certain duck species (such as the Mallard) that forage in shallow water, sometimes tipping forward to reach underwater food.

Decurved: curved downward (refers to the bill).

Dihedral: describes the upward-angled or V-shaped position (rather than flat or horizontal) in which certain birds hold their wings.

Diurnal: active during daylight hours.

Eclipse/Eclipse plumage: a brief period in late summer when waterfowl are molting from breeding plumage to nonbreeding plumage. Some ducks in eclipse plumage are very drab looking and cannot fly until their new feathers grow in.

Edge habitat: a place where two or more habitats overlap, such as an old meadow near woodlands. Edge habitat typically offers a rich diversity of birds.

Extinct: no longer existing.

Extirpated: no longer present in a given area (though still existing in others).

Eye line: the line over or through a bird's eye, often used as a field mark for identification. (See "Parts of a Bird," page 20.)

Eye patch: an area of feather (usually dark) surrounding a bird's eyes.

Eye-ring: a ring of color that encircles a bird's eye. A broken eye-ring is one that is not continuous, or does not completely encircle the eye.

Field mark: an obvious visual clue to a bird's identification, such as bill shape or plumage.

Flank: a bird's sides below the wings on either side of the belly. (See "Parts of a Bird," page 20.)

Fledgling: a young bird that has left the nest but may still be receiving care and feeding from a parent.

Flight call: a short, often distinctive call given by birds in flight.

Flock: a gathering of birds for purposes of feeding, resting, nesting, or migration. Winter feeding flocks of small woodland songbirds often contain several different species.

Forage: to look for food. Where and how birds forage can offer clues to their identities.

Habitat: the area or environment where a bird lives. Certain birds prefer specific types of habitat.

Hotspot: a location or habitat that is particularly good for bird watching on a regular basis.

Hybrid: the offspring produced from the mating of a male and a female from two distinct bird species.

Juvenal: plumage of a juvenile bird.

Juvenile: a bird that has not yet reached breeding age.

Life bird: a bird seen by a bird watcher for the first time, often recorded on a life list.

Life list: a record of all the birds a birder has seen at least once.

Local: refers to a species' abundance. A locally common species is one that is present in its appropriate habitat but not widespread and abundant.

Lores: the area between a bird's bill and its eyes. (See "Parts of a Bird," page 20.)

Malar: the area on the side of a bird's face below the bill and eye, sometimes referred to as the cheek.

Malar stripes: stripes in the malar or cheek area, often referred to as the mustache.

Mandible: the lower half of a bird's bill. The upper half is known as the maxilla.

Mantle: the upper back just behind the nape. (See "Parts of a Bird," page 20.)

Migrant: a bird that travels from one region to another in response to changes of season, breeding cycles, food availability, or extreme weather.

Mimic: a bird that imitates other birds' sounds and songs.

Molt: the periodic shedding of old feathers and their replacement by new ones.

Morph: a genetically fixed color variation within a species, such as the blue morph of the Snow Goose. (The term is correct only when both color variations occur in the same population. "Blue" morph Snow Geese breed side by side with white birds. The word *morph* is not applied to differently colored subspecies.)

Nape: the back of a bird's neck.

Neotropical migrant: refers to migratory birds of the New World, primarily those that travel seasonally between North, Central, and South America.

Nest parasite: a bird that lays its eggs in the nest of another bird or species, forcing that nesting bird or pair to raise the parasite's young. The most common North American example of a nest parasite is the Brown-headed Cowbird.

Nestling: a bird that has hatched from its egg but is still being cared for in the nest.

Nonbreeding plumage: the set of feathers worn during the fall and winter months. Many songbirds molt from breeding plumage into nonbreeding plumage in the fall. Sometimes called winter plumage or alternate plumage.

Nonmigratory: a bird that does not migrate with the change of seasons. Sometimes referred to as a resident bird.

Peeps: a generic term for confusingly similar small sandpipers.

Pelagic: birds of the ocean, rarely seen from land.

Pishing (or **spishing**): a sound made by bird watchers to attract curious birds into the open, made by repeating the sounds *spshhh* or *pshhh* through clenched teeth.

Plumage: collective reference to a bird's feathers, which change color and shape during seasonal molts. Breeding plumage is often more colorful than nonbreeding or winter plumage, worn during fall and winter.

Plumes: long showy feathers that are part of the high-breeding plumage of many herons and egrets. These feathers were once used to decorate women's hats. The collecting of these feathers decimated wading-bird populations.

Primary feathers (or **primaries**): the nine or more long flight feathers at the end of a bird's wing. (See "Parts of a Bird," page 20.)

Range: the area in which a bird may be seen during each season of the year. Breeding range refers to the area the species occupies during the breeding season; wintering range refers to the area occupied during the winter.

Raptors: birds of prey (hawks, eagles, owls).

Recurved: curved upward (refers to the bill).

Resident: a nonmigratory species that is present in the same region all year.

Rump patch: a patch of color located above the point at which a bird's tail connects to the body. (See "Parts of a Bird," page 20.)

Scapulars: the row of feathers lying just above a bird's folded wing; the lowest group of feathers on the mantle. (See "Parts of a Bird," page 20.)

Secondary feathers (or **secondaries**): the medium-length flight feathers located on the wing between the primaries and tertials. (See "Parts of a Bird," page 20.)

Shorebirds: refers to sandpipers, plovers, and related birds (but does not include herons, gulls, terns, and other birds found in coastal areas).

Skulker: a bird that does not make itself obvious, but keeps hidden in deep cover. Many sparrow species are referred to as skulkers.

Soaring: a flight style in which a bird holds its wings steady and flies without flapping. Red-tailed Hawks are experts at soaring.

Song: a complex series of sounds, with elaborate note patterns, usually associated with courtship or territoriality.

Song-flight: sometimes called flight songs. Performed by birds (usually males) during courtship, when they sing while flying high about their territories. Many grassland nesters perform song-flights, but some woodland species do too, such as the American Woodcock and a variety of warblers.

Speculum: the patch of inner secondary feathers on the wings of waterfowl. Mallards have a bright blue speculum.

Subadult: birds that are not yet adults but are more than one year old.

Supercilium: the area above a bird's eye, sometimes called the eyebrow. (See "Parts of a Bird," page 20.)

Tail spots: spots of contrasting color (usually white) on a bird's tail, often used as a field mark.

Territoriality: behavior associated with the aggressive defense of a particular area or territory.

Territory: the piece of habitat a bird claims for its own and defends against others of its species. Birds may be most territorial during the breeding season, but some birds (such as the Northern Mockingbird) will also defend food-rich winter territories.

Tertial feathers (or **tertials**): the innermost flight feathers on a bird's wing (the feathers closest to the bird's body), which form a stack atop the rear border of the folded wing. (See "Parts of a Bird," page 20.)

Underparts: the lower half of a bird (breast, belly, undertail), often used as a field mark.

Underwing: the bottom side of a wing.

Upperparts: the upper half of a bird (crown, back, top of tail), often used as a field mark.

Upperwing: the top side of a wing.

Vagrant: a bird that wanders far from its normal range.

Vent: the feathered area under the tail and below the legs, sometimes used as a field mark. (See "Parts of a Bird," page 20.)

Waders: herons, egrets, and related birds, including storks.

Wing bars: obvious areas of contrasting color, usually white, across the central portion ("shoulder") of a bird's wings. (See "Parts of a Bird," page 20.)

Wing lining: the inner portion of a bird's underwing.

Wingpit: the area on the underside of the wing, where it connects to the body.

Wintering grounds: the range over which a bird species spends the winter.

ACKNOWLEDGMENTS

The original idea for this book came out of the field trips our daughter Phoebe's elementary school class took to our southeastern Ohio farm each year. The fascination all kids have with birds is almost limitless, and though I know most of these kids will not become avid bird watchers, a few of them might, given the right encouragement, information, and tools. I decided to create a book that I wish I'd had as a young birder.

In 2008 that book, *The Young Birder's Guide to Birds of Eastern North America,* came out. It was gratifying to see the reception it got. One of the first people to buy a copy at my first promotional appearance was a woman who told me she was giving it to her grandson. I signed the book for her and she walked out of the room. About 15 minutes later I saw her get back in line with another copy in hand. When she finally reached my table she said, "You should have called this the *New Birder's Guide,* Bill! It's *perfect* for a new birder like me! It's great not just for young birders, but for young-at-heart birders, too!"

I realized she was right.

In 2012, Houghton Mifflin Harcourt published *The Young Birder's Guide to Birds of North America* (covering 300 commonly encountered birds from across the continent). And it was then that I began working on the idea for this book. I hope you like it, whether you're a young birder or merely young at heart.

This book is for those already interested enough in birds to want to know more. That was me, many years ago as a kid, and I owe a debt of gratitude to the adults who helped me along the path to becoming a bird watcher. In my younger days, Pat Murphy was a birding mentor to the Thompson family. My mom, Elsa, and much later, my dad, Bill Jr., my brother, Andy, and my sister, Laura, encouraged my interest in birds. Mom got us involved in the forays of the Brooks Bird Club in West Virginia, and today I owe much of my knowledge of birds to those kind and generous BBC members. After my parents started *Bird Watcher's*

Digest, I realized that I could watch birds for a living! This was quite a revelation, and I really never considered another career path.

This book is made immeasurably better by the beautiful illustrations by Julie Zickefoose. Julie also helped set me straight on a lot of natural history information, and she crafted many of the clever captions in the species profiles. I could not have completed this project without you, Jules.

In expanding the text of the original *Young Birder's Guide* to include the 100 additional western species, I was fortunate to have the illustration talents of Michael DiGiorgio, one of North America's best bird artists (and an excellent banjo picker). Mike's black-and-white illustrations are a great addition to the content.

Many wonderful photographers contributed their images to this book, an essential element for a birding guide since this activity is so visual.

Countless people have had a hand in molding me as a writer, editor, birder, and person. Naming everyone here is impossible, but I'd like to highlight just a few (in no particular order): Laura Thompson Fulton, Mary Beacom Bowers, Eirik A. T. Blom, Steve McCarthy, Shila Wilson, Chuck Bernstein, the Whipple Bird Club, and all of my friends in the Ohio Ornithological Society.

The faculty and staff of Salem-Liberty Elementary School permitted me to work with several classes of students on the original guide's development. I'd especially like to thank Phoebe's fourth-, fifth-, and sixth-grade classes at Salem-Liberty (by the time this book is published, Phoebe and her classmates will be finished with high school!): Mackayla Erb, Gregory Hill, Austin Klintworth, Drew Layton, Josh Lent, Hannah Lewis, Rikki Lockhart, Kristen Long, Brady Lowe, Alison Miller, Dana Moss, Tyson Niceswanger, Shannon O'Dell, Charity Roberts, Chelsey Schott, Amanda Seevers, Kaylee Sparks, Tara Thomas, Phoebe Thompson, Abbey Tornes, Dakota West, Paul Westfall, Jessica White, and their teachers: Mrs. Huck, Mrs. Baker, and Mrs. Biehl.

In my son Liam's fifth-grade class, my helpers for the update of the *Young Birder's Guide* included teachers Kelly Hendrix and Erica Schneider and students Jeffrey Bigley, Madison Binegar, Trae Dalton, Destiny Frye, Owen Gage, Hanna Grady, Nicholas Grosklos, Derik Hesson, Andrew Morganstern, Jesse Sloan, Kiersten Taylor, Morgan Wears, Kaylie Wittekind, Gage Treadway, Brenton Carpenter, Airetta Eales, Chiana Eddy, Barry Erb, Brooke Haas, Katelyn Jenks, Tyson Miller, Courtney Roberts, Tavion Rummer, Cody Sparks, Liam Thompson, Megan Tornes, Logan Wears, MacKenzie White, and Josh Zwick.

I'd like to thank the good folks at Houghton Mifflin Harcourt. My gratitude goes to Beth Burleigh Fuller, Katrina Kruse, Brian Moore, Taryn Roeder, Laney Everson, and a heaping helping of thanks to my (world-class) editor, Lisa A. White, for her guidance and friendship. My agent, Russell Galen, kept me firmly grounded in the early stages of this project, which he believed in from the very start. While I was writing this book, my colleagues at *Bird Watcher's Digest* kept things running and to them I owe my thanks, a nice celebratory staff lunch, some easy (and optional) birding around Marietta, and probably a few vacation days.

And last of all, thanks to you for reading this book. I'll see you out there with the birds.

—Bill Thompson III
November 2013
Whipple, Ohio

INDEX